Detection Methods for Cyanobacterial Toxins

Detection Methods for Cyanobacterial Toxins

Edited by

G. A. Codd
Department of Biological Sciences, University of Dundee, UK

T. M. Jefferies
School of Pharmacy and Pharmacology, University of Bath, UK

C. W. Keevil
Centre for Applied Microbiology and Research, Porton Down, Salisbury, UK

E. Potter
National Rivers Authority, Bristol, UK

THE ROYAL
SOCIETY OF
CHEMISTRY

The Proceedings of the First International Symposium on Detection Methods for Cyanobacterial (Blue-Green Algal) Toxins, held on 27–29 September 1993 at the University of Bath, UK

Cover photograph: a freshwater lake in Scotland showing shoreline accumulation of toxic cyanobacterial scum

Special Publication No. 149

ISBN 0-85186-961-0

A catalogue record for this book is available from the British Library

Published by The Royal Society of Chemistry,
Thomas Graham House, Science Park, Milton Road,
Cambridge CB4 4WF, UK

Printed in Great Britain by Bookcraft (Bath) Ltd

Preface

The toxic nature of cyanobacterial (blue-green algal) blooms (dense growths in waterbodies) and scums to animal and human health appears to have been long recognised in the folklore of aboriginal peoples in Australia and Canada. Reports of animal-, bird- and fish-poisoning incidents and of human health problems attributed to cyanobacteria have originated from several countries over recent decades. These incidents, plus the positive results of toxicity tests on cyanobacterial bloom samples collected from freshwaters in at least 35 countries, and from estuarine, coastal and marine waters, indicate a cosmopolitan occurrence of toxic factors in cyanobacterial blooms and scums.

The application of biological, toxicological, biochemical and physicochemical approaches and techniques to natural samples of cyanobacteria from diverse aquatic and terrestrial sources, and to pure laboratory cultures of cyanobacteria, has revealed at least 50 compounds which are toxic to vertebrates. The hazards presented by cyanobacterial toxins to human health are beginning to be understood as knowledge of their toxicity at acute and sub-acute exposures in animals increases, as the recognition of human health problems attributed to the toxins develops, and as data emerge on the types and concentrations of the toxins which may occur in waterbodies.

Research on the fundamental properties of the toxins, on monitoring their occurrence in natural environments, and in controlled waters used for recreation, aquaculture and potable supply, is dependent upon adequate methods of detection and quantification. For several years, the principle method for cyanobacterial toxicity testing was the intraperitoneal mouse bioassay. This method is reproducible and distinguishes between classes of cyanobacterial toxins, e.g. hepatotoxins and different types of neurotoxins. However, limitations in specificity, sensitivity, availability and humane considerations argue for the development of additional methods.

Research and development on the analysis of cyanobacterial toxins is in progress in several of the countries affected by toxic cyanobacterial blooms. The institutions in which this work is in progress, or of interest, include universities, national public health service laboratories and environmental protection agencies, and drinking water undertakings in the public and private sectors. The need of these various bodies to know what types of cyanobacterial toxins are present in an aqueous environment, how abundant they are and where they are located, may vary. Nevertheless, in order to reduce duplication of effort, encourage comparability and to facilitate the development and acceptance of methods, increased contact and exchange between the various user-groups is necessary.

With these aims in mind, the First International Symposium on Detection Methods for Cyanobacterial (Blue-Green Algal) Toxins was held at the University of Bath in September 1993. The Symposium was organised by the Department of the Environment's Standing Committee of Analysts Algal Toxins Panel as part of its remit to consider standard methodology for the water industry and other water-users and regulators. The Symposium was supported by the following UK institutions: The Royal Society of Chemistry, The National Rivers Authority and The Drinking Water Inspectorate. 120 participants from 17 countries attended the meeting.

These Proceedings include all of the written contributions received, which are based on oral and poster presentations. Most of the contributions focus on the development and merits of methods for the detection and quantification of cyanobacterial toxins. These may be assessed in terms of requirements of sensitivity, specificity and other characteristics as indicated in accompanying contributions on the health significance of the toxins and environmental health risk studies. The Proceedings indicate that a range of biological and physicochemical methods for cyanobacterial toxin detection and analysis is under development. The merits of individual methods and the combination of methods in a multistage programme for the screening, quantification and confirmation of cyanobacterial toxins in various matrices can be expected to emerge with the continuing investigations into cyanobacterial toxins in basic research and the protection of water quality.

G.A. Codd, University of Dundee
T.M. Jefferies, University of Bath
C.W. Keevil, Centre for Applied Microbiology and Research, Porton Down
E. Potter, National Rivers Authority, Bristol

October 1994

Contents

Oral Presentations

Poster Presentations

Oral Presentations

Health Problems from Exposure to Cyanobacteria and Proposed Safety Guidelines for Drinking and Recreational Water

Ian R. Falconer

DEPUTY VICE-CHANCELLOR (ACADEMIC), UNIVERSITY OF ADELAIDE,
ADELAIDE, SOUTH AUSTRALIA 5005

Poisoning from the drinking of 'bad water' is part of human folklore, and we naturally avoid smelly, discoloured or foul-tasting water to drink. So do wild animals, birds and domestic livestock. Nevertheless there are many reports of stock deaths attributed to drinking water contaminated with cyanobacteria, particularly from the U.S.A., South Africa and Australia. In my experience livestock will only drink water containing cyanobacteria when there is no other choice - the supply is contaminated and the stock cannot wade out into clean water, or move to an alternative supply. Under hot conditions all livestock are compelled to drink to survive, and with hot conditions in temperate latitudes cyanobacterial water blooms may flourish. Rivers and reservoirs frequently contain sufficient nutrients for cyanobacterial abundance, and the weather and other water conditions determine the extent of proliferation of the organisms. The four species responsible for most stock deaths are *Microcystis aeruginosa*, *Anabaena circinalis*, *Nodularia spumigena* and *Aphanizomenon flos-aquae*, in descending order of cases[1].

Microcystis and *Nodularia* Toxicity

The toxins isolated from *Microcystis* form a family of cyclic peptides (microcystins) of seven amino acids, whose chemistry will be extensively described elsewhere in this volume. *Nodularia* contains a toxic cyclic peptide (nodularin) of five amino acids, and both the seven and five amino acid peptides act as toxins in a biologically identical manner[1].

The primary effect on health is toxicity to liver cells (hepatocytes), as a consequence of selective transport mechanisms which concentrate the peptide toxins from the blood into the liver. As with all toxic material consumed as food or drinking water, the toxins need to traverse the gut lining in order to enter the bloodstream. The cells lining the small intestine have transport mechanisms for bile-acid uptake, and these carry the peptide toxins across the gut cells to the hepatic portal vein[2]. From this blood supply the toxins are distributed to the liver lobules, and are taken up by the hepatocytes which line the liver sinusoids.

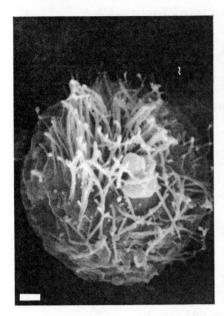 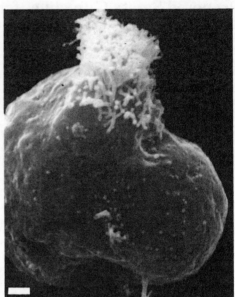

Figure 1. Isolated gut cells. Left, control cell, spherical
with microvilli; right, toxin deformed cell with large bleb.
Bar 1.0μm.

Cells from the gut lining and from the liver are damaged,
the damage being visible as a physical distortion of the cells
(Figure 1) which can be seen under conventional or scanning
electron microscopy. The underlying biochemical injury is a
powerful inhibition of specific phosphatase enzymes resulting
in hyperphosphorylation of proteins, which is exhibited by a
breakdown of intermediate filaments of the cell cytoskeleton
and a retraction of actin microfilaments[3]. The cell distortion
is such that the organizational structure of the liver itself
falls apart, and the animal bleeds into its own liver causing
death. At lower doses of toxins, enteritis and hepatitis are
seen in man and animals, shortly after the drinking of toxic
Microcystis[4].

Secondary effects on health are similar to those of
infections of the liver or toxic liver damage, and include
liver insufficiency, photosensitization and jaundice. The
particular biochemical inhibition caused by these peptides is
identical to that produced by the sponge toxin okadaic acid,
and both have been experimentally demonstrated to be tumour
promoters in the liver[5,6]. The potential public health impact
of these toxic cyanobacterial peptides in drinking water is
receiving considerable attention at present, and is discussed
later.

Anabaena and *Aphanizomenon* Toxicity

The second most destructive cyanobacterium to livestock is *Anabaena circinalis*, which occurs in rivers as well as reservoirs. The toxins isolated from this organism include the microcystins, but their presence is normally hidden by the rapid death of the animal from neurotoxicity, before liver injury can take effect. In Australia we have been puzzled for some years as to the neurotoxins present, because we have been unable to find the two toxins already well described. These were both first isolated from Anabaena strains in the U.S.A., and have been characterised by Carmichael and colleagues[1]. One is anatoxin-a, a neuromuscular blocking alkaloid, and the other anatoxin-a(s) which is an organophosphate acting as an anticholinesterase.

Work earlier this year, and later to be described by Steffensen in this volume, has shown a range of paralytic shellfish poisons – the alkaloids saxitoxin, neosaxitoxin, gonyautoxin etc. in a series of samples of Australian *Anabaena*. These alkaloids block sodium channels in nerves, hence causing paralysis. This too has implication for public health, particularly as the samples come from a widespread series of drinking water sources. *Aphanizomenon* collected in New Hampshire in the U.S.A. also contains saxitoxins.

Human Toxicity

While it is relatively easy to attribute stock deaths to cyanobacterial poisoning, as a result of observing the presence of dead stock adjacent to water contaminated with a toxic cyanobacterial bloom, and the identification of cyanobacteria in the gut of poisoned animals, it is far harder to study human injury which can be clearly ascribed to cyanobacterial poisoning. In the following paper by Hunter, some of the problems of human epidemiology will be explored.

The nature of the risks to human populations can, however, be predicted from animal observations, and then tested by clinical observations. As far as I am aware, no human deaths have been recorded as a verified consequence of cyanobacterial poisoning. However, the absence of deaths from acute poisoning is more a tribute to the skill of the medical profession than an indicator of toxicity, in several specific instances of human poisoning in the U.K. and Australia. To attribute deaths from liver cancer in the population to cyanobacterial tumour promotion, through chronic consumption of contaminated water, is difficult. It will require detailed epidemiology in areas of high risk of cyanobacterial toxins in water supplies. In areas of China in which liver cancer has a high incidence, epidemiological studies are in progress to explore possible links to diet, including cyanobacterially contaminated drinking water.

Toxin in Drinking Water

There are two distinct human health problems occurring as a consequence of cyanobacterial contamination of water. The first is from toxin in drinking water supplies. Almost all the reported cases of sickness of large numbers of people which have been attributed to cyanobacteria, have followed the lysis of water blooms of the organisms in water supply reservoirs. Lysis can occur naturally, or as a result of copper sulphate dosing of a reservoir, or after chlorination of water for drinking. Many drinking water supplies are simply chlorinated water from rivers or lakes, with no water treatment plant to filter off organisms or absorb toxins. Such a supply at Palm Island, off the Queensland coast of Australia, was dosed with copper sulphate to remove a persistent cyanobacterial bloom. Within a week an epidemic of severe hepatoenteritis caused 136 individuals to be hospitalised, some children were critically ill and only survived through intravenous therapy[7]. Investigation of the dam resulted in the identification of a filamentous cyanobacterium *Cylindrospermopsis raciborskii* as the predominant organism, and toxicity tests on extracts showed a general cytotoxin present causing widespread organ injury and thrombosis. This toxin has since been identified as an alkaloid and named cylindrospermopsin[8,9]. Another similar organism, *Schizothrix calcicola*, caused a gastroenteritis epidemic in a community in the U.S.A., but up to the present the toxin has not been identified.

As a result of our monitoring of the toxic *Microcystis* present in the drinking water supply reservoir of Armidale, a small isolated city in NSW, Australia, it was possible to retrospectively identify a time of peak risk to the population. During the summer of 1981 the *Microcystis* cell number increased in the reservoir, until it became a surface bloom accumulating along the dam wall adjacent to the water offtake tower for about three weeks. At this time the local authority spread copper sulphate over the reservoir at 1 mg/litre in the top metre of water, which caused total lysis of visible cells and scum within a week. By negotiation with the Regional Pathology Service, we obtained the clinical liver function data from blood samples and clinical records from all patients tested over a wide geographical area during a 15 week period. The patients were sorted by place of residence (hence drinking water supply), and date of testing. We analysed the data for four potential enzyme markers of liver injury, and also read each clinical report to evaluate alcoholism, infectious hepatitis or any clinical condition likely to affect the data. The results showed that a statistically significant rise in a marker enzyme for toxic liver injury, γ-glutamyl transpeptidase (GGT), was apparent in the patients drinking from the contaminated reservoir, only during the 5 week period of peak bloom and lysis, and only in patients on the affected water supply[10].

The treatment plant processing the water from this contaminated reservoir was a conventional high quality unit, employing pre-chlorination, pH adjustment, flocculation with alum, gravity sedimentation, rapid sand filtration, post-chlorination, fluoridation and reticulation. The plant had one major disadvantage of location, being about 15 km from and 300 m lower than the storage reservoir. Hence any cyanobacteria drawn into the intake would be likely to be lysed or injured in the pipeline and pressure reduction valves and the toxins consequently freed into the water before entering the treatment plant.

While there is no publicly available data for measurements of cyanobacterial toxin content of domestic water supplies at the present, there exists data in the U.S.A., U.K. and Australia showing the presence of toxins or toxicity in tap water. It is an indication of the present early stage of development of toxin assays, that the U.S.A. study used an Enzyme-Linked Immunosorbent Assay, the U.K. study High Performance Liquid Chromatography and the Australian study mouse bioassay.

Cyanobacteria in Recreational Water

Recreational exposure to cyanobacteria is a quite separate potential hazard to the population. It is voluntary exposure, unlike drinking water supplies. Warning signs can be erected at modest cost to local authorities, and if swimmers use the signs to dry their towels and bathe regardless, it is at their own risk. It is, however, necessary to evaluate the nature and extent of the risk, and to provide effective advice.

By collection of clinical data from cases of injury through recreational exposure to cyanobacteria, a pattern of response can be found. Probably the most commonly reported effect is skin irritation during and after watersports, or after showering in the water. Most acutely the skin forms blisters in sensitive areas such as the lips. The worst offender is a marine filamentous cyanobacterium in tropical waters - *Lyngbya majuscula*, which when caught under the swimsuit will cause severe blistering and deep desquamation of skin in very sensitive areas[11].

The skin reactions may be allergic, or due to toxin penetration of the skin, and both are evident in different cases. Conjunctivitis, hay-fever, and asthma symptoms are also recorded from bathers who have been exposed to toxic water blooms. One water skier in Australia exposed to a spray of *Anabaena* probably containing paralytic shellfish poisons, complained of difficulty in breathing of sufficient severity to seek medical aid.

Oral consumption of toxic cyanobacteria during recreation is relatively rare, though the more energetic of water contact sports such as falling off a water ski, playing water polo or being ducked while windsurfing, canoeing or sailing, can result

in swallowing water. The most severe case of recreational
exposure was in the U.K., where pneumonia, hepatoenteritis and
skin blistering were reported in army trainees compelled to
swim and roll canoes in a lake with a bloom of toxic
Microcystis[4].

Water Safety Guidelines

Because of the recent worldwide recognition of potential
health problems caused by toxic cyanobacteria, it is necessary
to derive water safety guidelines for toxins in drinking
water, and also guidelines for safety of recreational waters.
At present these guidelines have to be derived from mixed
sources of data, and are inevitably partial. The best
established toxicity data is for the cyclic peptide toxins,
which have been subject to extensive animal testing. Using
data for intra-peritoneal toxicity in mice, for chronic oral
toxicity in mice, and particularly using new data for sub-
chronic oral toxicity in pigs carried out in my own
laboratory, it is possible to derive "lowest observed adverse
effect levels"[12]. By application of a series of safety factors
used in occupational health and drug evaluation, these can be
used for calculating guideline levels for safe exposure (Table
1).

From these calculations, a reasonable safety guideline
for sub-chronic exposure to peptide toxins in drinking water
is 1 µg *Microcystis* toxin/litre.

At present further work is needed to derive similar
guidelines for neurotoxins, though extrapolation from adopted
safety guidelines for shellfish toxins will reduce the need
for experimental data. The U.S. guideline values for
paralytic shellfish poisons in shellfish meat for human
consumption are specified as safe below 80 µg/100 g tissue[13].
If 100 g of oyster meat can be imagined as a day's intake, it
equates to 2 litres of water. Thus perhaps 40 µg/litre of
shellfish poisons is acceptable in a water supply, though I
personally would put in a further safety factor of 10 on the
basis of regularity of water consumption compared to
shellfish!

Table 1. Safety factors in the determination of water
          guidelines for microcystins

Subchronic data to lifetime risk	–	10
Pig data to human risk	–	10
Intra-human population variation	–	10
Tumour promotion risk	–	10
Overall safety factor		10,000

Basic data – subchronic pig exposure. Lowest observable
effect level 280 µg toxins/Kg bodyweight/day.
Intake assumption 2 litres water per day by 60 Kg adult,
therefore 16.8 mg/day in 2 litres = 8.4 mg/litre
    ÷ 10,000 = 0.84 µg/litre – <u>approximately 1 µg/litre</u>
can be regarded as maximum safe concentration.

Recreational waters provide a real problem in defining safety guidelines. In a single water body there is likely to be a very wide variation in cell numbers of cyanobacteria per volume of water, depending on the location of sampling. In shallow bays on the downwind side of a lake during a cyanobacterial bloom, cells in the top 10 cm of water can often be in concentrations of 10^6/ml and above. Out in turbulent water offshore the cell numbers may be 10,000 cells/ml. Cell numbers will change markedly with weather and day or night, as the cyanobacteria move up or down in the water. Toxicity of cells also varies, from undetectable toxicity (or almost so) to 10 mg of dry algae being lethally toxic to 1 kg mice.

Because it is necessary to provide some workable guidelines to local authorities, and only the most basic techniques are likely to be available for monitoring recreational waters, cell or colony counting is recommended. In Australia we have recommended 20,000 cells/ml in recreational waters as a maximum safe level, based on sampling the top metre of open water. At this cell concentration, sustained good weather conditions will allow scums of concentrated cyanobacteria to form which will then present a significant hazard if the scums are toxic. I think we would all agree that swallowing or bathing in a toxic *Microcystis* scum is a real health hazard!

Conclusion

Cyanobacterial water blooms are becoming an increasing problem in many parts of the world. They occur in drinking water supply reservoirs, rivers used as drinking water sources, and in recreational waters. There is clear evidence of injury to people and livestock from consuming cyanobacterial toxins from drinking water supplies or by accident during recreation. A further potential hazard of stimulation of cancer growth has been identified. Water safety guidelines are proposed, with 1 µg/l cyanobacterial peptide toxins as a maximum concentration in drinking water, and 20,000 cells/ml of cyanobacteria as a maximum for safe use of recreational waters.

REFERENCES

1. W.W. Carmichael and I.R. Falconer. In: Algal Toxins in Seafood and Drinking Water, (ed. I.R. Falconer), Academic Press, London, 1993.

2. I.R. Falconer, M. Dornbusch, G. Moran and S.K. Yeung. <u>Toxicon</u>, 1992, <u>30</u>, 790-793.

3. I.R. Falconer and D.S.K. Yeung. <u>Chem. Biol. Interact</u>. 1992, <u>81</u>, 181-196.

4. P.C. Turner, A.J. Gammie, K. Hollinrake and G.A. Codd. <u>Br. Med. J.</u> 1990, <u>300</u>, 1440-1441.

5.   I.R. Falconer. Environ. Toxicol. Water Qual. 1991, 6,
     177-184.

6.   R. Nishiwaki-Matsushima, T. Ohta, S. Nishiwaki, M.
     Suganuma, K. Kohyama, T. Isikawa, W.W. Carmichael and H.
     Fujiki. J Cancer Res. Clin. Oncol. 1992, 118, 420-424.

7.   A.T.C. Bourke, R.B. Hawes, A. Neilson and N.D. Stallman.
     Toxicon (suppl.) 1983, 3, 45-48.

8.   P.R. Hawkins, M.T.C. Runnegar, A.R.B. Jackson and
     I.R.Falconer. Appl. Environ. Microbiol. 1985, 50, 1292-
     1295.

9.   I. Ohtani, R.E. Moore and M.T.C. Runnegar. Amer. Chem.
     Soc. 1992, 114.

10. I.R. Falconer, A.M. Beresford and M.T.C. Runnegar. Med.
     J. Aust. 1983, 1, 511-514.

11. A.H. Banner. Hawaii Med. J. 1959, 19, 35-36.

12. I.R. Falconer, M.D. Burch, D.A. Steffensen, A. Choice and
     O.R. Coverdale. Environ. Toxicol. Water Qual. 1994, 9,
     131-139.

13. I.R. Falconer, (Ed), Algal Toxins in Seafood and Drinking
     Water,Academic Press, London, 1993.

An Epidemiological Critique of Reports of Human Illness Associated with Cyanobacteria

P. R. Hunter

PUBLIC HEALTH LABORATORY, COUNTESS OF CHESTER HEALTH PARK,
LIVERPOOL ROAD, CHESTER CH2 3UL, UK

1 INTRODUCTION

Several of the other articles included in this book give a very good indication of the variety and nature of the toxins that are produced by the cyanobacteria. The potential hazards associated with these toxins have been adequately demonstrated in animals both by experiment and by observations of accidental poisoning. However, the identification of hazard is only the first step towards assessing whether cyanobacteria pose a risk to public health. Proving that cyanobacteria are indeed a risk to human health requires detailed epidemiological study.

Most medical students are aware that for an organism to be proven to cause disease in humans that organism should be shown to satisfy Koch's postulates.[1] Koch's postulates were:-

1. the organism is regularly found in the lesions of the disease

2. it can be isolated in pure culture on artificial media

3. inoculation of this culture produces a similar disease in experimental animals

4. the organism can be recovered from the lesions in these animals.

Unfortunately, Koch's postulates are of no help in assessing the health implications of the cyanobacteria. Many of the diseases that may be associated with cyanobacterial poisoning are not a direct result of infection but are mediated by toxins. The organism is, therefore, only rarely found in clinical specimens. Furthermore, there is no single disease, lesion or symptom complex that can be linked with cyanobacterial contact. Rather a variety of symptom complexes have at some time been claimed to be due to cyanobacteria or their toxins.

In 1965 Bradford-Hill suggested nine epidemiological criteria to be used in assessing whether an environmental factor was associated with human disease.[2] These factors are:

1.   Strength of association
2.   Consistency
3.   Specificity of association
4.   Temporality
5.   Biological gradient
6.   Plausibility
7.   Coherence
8.   Experiment
9.   Analogy

In an attempt to assess the risk of cyanobacterial poisoning to human health, each of these criteria will be discussed in turn. Whether or not reported incidents of cyanobacterial disease in humans satisfy that criterion will then be discussed.

Episodes of human illness that have been associated with cyanobacteria can be classified as to how the affected individuals were exposed to the cyanobacteria or toxins. This contact may have been by recreational contact with algal blooms on surface water, by the consumption of fish, by airborne spread and by contamination of potable water supplies. This chapter will only consider reports of illness that have implicated potable water supplies. More detailed discussion of illnesses associated with other routes of contact have been discussed elsewhere.[3]

2 CRITIQUE OF REPORTS OF HUMAN ILLNESS IMPLICATING CYANOBACTERIAL CONTAMINATION OF POTABLE WATER SUPPLIES

Each of Bradford-Hill's criteria will now be discussed in turn.

Strength of association

The strength of association criterion relates to the statistical significance of disease rates in exposed and non-exposed populations. The degree of significance is usually measured by the Chi-squared or Fisher's exact probability test.

Relatively few of the reported outbreaks give adequate information to determine whether of not they satisfy this criterion. This is not surprising as many of the reports date back several decades, when analytical epidemiology was not widely practiced.

Falconer, Beresford and Runnegar reported a retrospective study of liver enzyme results in the city of Armidale, New South Wales.[4] This study took place during the time that one of the reservoirs, the Malpas Dam, developed an algal bloom due to *Microcystis*. Liver enzyme results from people living in the supply area of the Malpas Dam were compared with people living in the surrounding countryside who received their water from other sources. They found that a biochemical marker of acute hepatitis, γ glutamyl-transferase (GGT), was more likely to be elevated in the population receiving their drinking water from the Malpas Dam at the time of the bloom. This was highly statistically significant ($p<0.005$). Despite the biochemical evidence of sub-clinical hepatitis in the Malpas Dam population, there was no difference in the incidence of clinically obvious liver disease between the two populations. Although suggestive of an association this study does not prove the hypothesis that the algal bloom was responsible for the elevation of GGT. Criticisms of the report could include the fact that it was a retrospective study that did not attempt to control the two populations in any way. Indeed, there were differences in age between the two populations. Furthermore no epidemiological studies were reported to link elevations of GGT with dose of water consumption.

Another paper reported the epidemiological investigation of a seasonal outbreak of diarrhoea affecting US forces personnel on the Clark Air Base in the Philippines.[5] These epidemics occurred during the hottest season of the year (March to July) and affected up to 6,000 of the 36,000 Americans on the base. The epidemics stopped each year when the heavy rains came. All cases suffered from diarrhoea, and many had abdominal pain, nausea, weakness, and fatigue. One fifth of cases had illness lasting for more than two weeks, and in these malabsorption occurred. There was a significant excess of cases in people working on those parts of the base that tended to be supplied with river rather than well water ($p = 0.002$). Although the epidemiology was inconclusive, the authors suggested a temperature dependant agent, possibly blue-green algae or bacteriophage, in the river water from which most of the drinking water was taken.

Consistency

This refers to whether the disease has been observed to be linked to the proposed environmental agent by a variety of different workers at different times and places. Clearly this criterion has been satisfied by the number of reports referred to in this chapter.

Specificity of association

This criterion relates to whether a particular type of exposure leads to a particular disease. As can be seen from the range of clinical syndromes described, the criterion of specificity is not satisfied. This is not surprising in view of the diversity of toxins produced.

Although the criterion of specificity of association cannot completely be satisfied, it must be noted that it could be said to have been satisfied in relation to gastroenteritis. In addition to the outbreaks of gastroenteritis at the Clark Air Base described above,[5] there have been several other outbreaks of gastroenteritis associated with cyanobacterial contamination of potable water supplies.

Zilberg reported a study of childhood hospital admissions due to gastroenteritis in Salisbury, Rhodesia.[6] This study covered the five years from July 1960 to June 1965. Whilst complaining that a laboratory diagnosis was not made in the majority of cases, Zilberg noted a sharp annual increase in admissions during winter from an area supplied by a reservoir that regularly suffered from algal blooms. Populations supplied by reservoirs that were not affected by algal blooms did not show this seasonal peak. The peak of gastroenteritis corresponded with the time that the algae on the affected reservoir were dying. Zilberg postulated that the peak of gastroenteritis was caused by toxins released by the algae as they died.

During 1975 an outbreak of gastroenteritis affected an estimated 5,000 people supplied by a single reservoir in Sewickley, Pa.[7] Although clinically mild this represented an attack rate of greater than 50% of the population supplied from this reservoir. No microbial pathogen was implicated, but on examination of the reservoir, the remains of a bloom of *Schizothrix calicola* was obvious. Subsequent testing of this bloom in following years showed some mouse toxicity and endotoxin production.[8]

Other outbreaks of gastroenteritis linked to cyanobacterial blooms affecting potable water supplies are reviewed elsewhere.[9]

Temporality

Temporality refers to the time relationship between exposure and the onset of disease. It is clearly important to distinguish between illness caused by contact with an environmental agent and the possibility that patients suffering from the disease seek out contact with the agent.

A recurring theme with several reported outbreaks is that disease followed the death of a cyanobacterial bloom. Such cell deaths were either natural or the result of dosing water reservoirs with algicides. This was certainly the case in the Palm Island Mystery Disease. In November 1979, an outbreak of disease affected 139 children and 10 adults of aboriginal descent in the Palm Island Community, Queensland, Australia.[10] Although later termed hepatoenteritis, this name is misleading as many organ systems seem to have been affected and the enteritis was only seen in a minority of patients. The illness began as an acute hepatitis with malaise, anorexia, vomiting & tender hepatomegally. On presentation 74% had glycosuria, 89% proteinuria, 20% haematuria, and 53% had ketonuria. During the next few days 82% developed acidosis and hypokalaemia, often with severely abnormal serum electrolytes. Later 39% developed diarrhoea of which 92% passed frank blood. All cases survived, recovery taking between 4 and 26 days. Laboratory investigations of a possible cause were singularly unhelpful. Epidemiological investigations were not very helpful, although it was noted that no illness was reported amongst people whose water supply was not connected to the general supply.[11] It was noted that the Solomon Dam which provided most of the water for the community had developed a heavy bloom in October, such that the local drinking water was discoloured and had a disagreeable odour and taste. The dam had been treated with copper sulphate five days before the start of the epidemic to kill the algae. Subsequent investigations of blooms on the Solomon Dam found that one of the main components of the bloom, *Cylindrospermopsis raciborskii*, causes major liver damage in mice, as well as effecting the kidneys, adrenals, intestines, and lungs.[12]

Perhaps the best example of temporality was an outbreak of endotoxaemia. Patients attending a private dialysis clinic near Washington DC developed pyogenic reactions during a four week period in July and August 1974.[13] The authors documented 49 reactions in 23 patients characterised by chills, fever, myalgia, nausea and vomiting, and hypotension. High levels of endotoxin were detected in potable water supply by the Limulus lysate test. On further investigation of the water supply it was found that the epidemic corresponded with a period of high blue-green algal bloom counts. When the algal counts in the supply reservoir declined endotoxin disappeared from the water supply and the outbreak stopped.

Biological gradient

The criterion of biological gradient suggests that there should be a relationship between the amount of exposure to an environmental agent and the severity of disease or the probability of developing disease. As far as I am aware, no epidemiological study so far published has demonstrated a dose response curve between exposure to cyanobacterial contaminated potable water and human disease. This is a weakness in the current literature and needs to be addressed in future studies.

Plausibility

When considering the plausibility of any association we are asking whether the proposed link seems likely, given our current state of knowledge. Whether or not one accepts the plausibility of the reports referred to in this chapter is, in part, a subjective assessment on the part of the reader. Nevertheless, I would suggest that the chapter by Falconer, elsewhere in this book, strongly supports the criterion of plausibility.

Coherence

For an association to be coherent, it must be compatible with what is know about the biology of the disease. Considerable advances have been made in recent years on the possible pathophysiology of cyanobacterial disease. This has been particularly true for the microcystins.

Epidemiological research over many years in China has shown that the incidence of primary liver cancer is linked with the source of drinking water. In particular, rural populations drinking surface water are more likely to suffer from hepatic cancer than those drinking well water.[14] Recent work by Carmichael, reported elsewhere in this book, has suggested that this association with potable water is due to cyanobacterial contamination of water supplies. The coherence of this suggestion is supported by recent work on the biological effects of the microcystins.

Falconer and colleagues showed that an extract of *Microcystis* fed to mice in drinking water acted as a potent skin tumour promoter in association with topical applications of dimethyl-benzanthracene.[15] They reported that the mice receiving both cyanobacterial extract and dmb had more tumours than those receiving dmb alone. Furthermore, those mice had larger and more vascular tumours. No tumours were seen in mice receiving cyanobacterial extracts without dmb. More definitive animal evidence has come from Japanese work which concluded that microcystin is the most potent liver carcinogen yet described.[16]

Recent work by MacKintosh *et al.* has shown that microcystins are potent inhibitors of protein phosphatases type 1 and type 2A.[17] Microcystins bind covalently to these enzymes and cause a long term inhibition of their activity. Protein phosphatases have a major role in many cell processes such as cell growth and division, gene transcription, protein synthesis, etc. Interference with these processes can certainly explain many of the adverse effects that have been attributed to the microcystins.

The biochemical effects of microcystin may also explain possible links between algal blooms and human birth defects. Collins *et al.* reported the results of an epidemiological investigation into a high birth defect rate that suggested a link to two drinking water reservoirs.[18] Subsequent laboratory studies found mutagenicity in the reservoir water in the Ames/Salmonella test and in a colony of Sprague-Dawley rats. This mutagenicity was present only at the time of algal blooms, predominantly due to *Oscillatoria subbrevis*.

Experiment

To satisfy the criterion of experiment one should attempt to demonstrate a reduction in the incidence of a disease by taking action to reduce contact with the environmental agent.

As yet this criterion has not been satisfied. Most of the incidents described in this chapter have been relatively short-lived. Therefore, there is little ability to intervene in most outbreaks.

Analogy

This criterion requires that similar diseases are caused by related environmental agents. Although not affecting water supplies, there are related diseases associated with paralytic and diarrhetic shell fish poisoning. These diseases are caused by consuming shell-fish that have absorbed toxins from marine dino-flagellates.[19]

3 CONCLUSIONS

As discussed above, many of the reports of presumed cyanobacterial poisoning of humans are open to criticism on epidemiological grounds. No study has satisfied all of Bradford Hill's criteria. In particular, case definitions were poor and very few studies reported any analytical epidemiology.

Nevertheless, the weight of evidence from animal studies and from the human illnesses that have been documented provides strong evidence that cyanobacteria are a hazard to human health. This evidence is further supported by the observations that examination of the effects of implicated algal blooms in animals frequently mirrored the disease in humans.

Even with the knowledge of the hazards described above, the risks of cyanobacterial contamination of potable water supplies are still far from clear. It is impossible from the studies that have been reported to get any idea how frequent such illness may be in the U.K. or elsewhere.

There is a clear need to develop appropriate analytical methods for detecting cyanobacterial toxins in water supplies. The information gained from such analytical work must also be linked to appropriate and detailed analytical epidemiological and microbiological investigations of exposed populations.

REFERENCES

1. C.A. Mims, J.H.L Playfair, I.M. Roitt, D. Wakelin and R. Williams, 'Medical Microbiology', Mosby Europe Ltd, London, 1993, Chapter 10, p. 10.6.

2. A. Bradford Hill, <u>Proc. Roy. Soc. Med.</u>, 1965, <u>58</u>, 295.

3. P.R. Hunter, <u>PHLS Microbiol. Digest.</u>, 1991, <u>8</u>, 96.

4.   I.R. Falconer, A.M. Beresford and M.T.C. Runnegar, <u>Med.</u>
      <u>J. Aust.</u>, 1983, <u>1</u>, 511.

5.   A.G. Dean and T.C. Jones, <u>Am. J. Epidemiol.</u>, 1972, <u>95</u>,
      111.

6.   B. Zilberg, <u>Cent. Afr. J. Med.</u>, 1966, <u>12</u>, 164.

7.   E.C. Lippy and J. Erb, <u>J. Amer. Water Works Ass.</u>, 1976,
      <u>76</u>, 606.

8.   W.W. Carmichael, C.L.A. Jones, N.A. Mahmood and W.C.
      Theiss, <u>CRC Crit. Rev. Envir. Contr.</u>, 1985, <u>15</u>, 275.

9.   M. Schwimmer and D. Schwimmer, 'Algae and Man', D.F.
      Jackson (ed), Plenum Publishing Corporation, New York,
      1964, p. 368.

10.   S. Byth, <u>Med. J. Aust.</u>, 1980, <u>2</u>, 40.

11.   A.T.C. Bourke, R.B. Hawes, A. Neilson and N.D. Stallman,
      Toxicon Suppl 1983; 3: 45.

12.   P.R. Hawkins, M.T.C. Runnegar, A.R.B. Jackson and I.R.
      Falconer, <u>Appl. Envir. Microbiol.</u>, 1985, <u>50</u>, 1292.

13.   S.H. Hindman, M.S. Favero, L.A. Carson, N.J. Petersen,
      L.B. Schonberger and J.T. Solano, <u>Lancet</u> 1975, <u>2</u>, 732.

14.   S.-Z. Yu, 'Primary Liver Cancer', Z.Y. Tang, M.C. Wu and
      S.S. Xia (eds), Springer, Berlin, 1989, p. 30.

15.   I.R. Falconer and T.H. Buckley. <u>Med. J. Aust.</u>, 1989, <u>150</u>,
      351.

16.   R. Nishiwaki-Matsushima, T. Ohta, S. Nishiwaki, M.
      Sugawama, M. Kohyama, T. Ishikawa, W.W. Carmichael and H.
      Fujiki, <u>J. Cancer Res. Clin. Oncol.</u>, 1992, <u>118</u>, 420.

17.   C. MacKintosh, K.A. Beattie, S. Klumpp, P. Cohen and G.A.
      Codd, <u>FEBS Lett.</u>, 1990, <u>264</u>, 187.

18.   M.D. Collins, C.S. Gowans, F. Garro, D. Estervig, T.
      Swanson, 'Algal toxins and health', W.W. Carmichael (ed),
      Plenum Press, New York, 1981, p. 271.

19.   A.C. Scoging, <u>Comm. Dis. Rep.</u>, 1991, <u>1</u>, R117.

Cyclic Peptide Hepatotoxins from Fresh Water Cyanobacteria Water Blooms Collected in the River Dnieper Reservoirs and Other Water Bodies from the European Part of Russia

V. M. Tchernajenko

DEPARTMENT OF RADIATION AND MOLECULAR BIOPHYSICS, ST. PETERSBURG NUCLEAR PHYSICS INSTITUTE, RUSSIAN ACADEMY OF SCIENCES, GATCHINA 188 350, ST. PETERSBURG, RUSSIA

1 INTRODUCTION

Occurrence of toxic cyanobacteria water blooms have been documented in many parts of the world beginning from the end of the past century to present days [1,2,3]. It is well known that the main toxic bloom-forming cyanobacteria include the genera *Anabaena, Aphanizomenon, Microcystis, Nodularia* and *Oscillatoria* [4]. Two groups of toxins, the neurotoxic anatoxins and/or hepatotoxic cyclic peptide microcystins, have been identified in species and strains from these genera [5].

The first water blooms in the former USSR that were toxic were found in reservoirs of the River Dnieper. These reservoirs were constructed for electric power production [6,7]. The organisms responsible for toxicity and the toxins' structure were not investigated in detail. More recently Sivonen et al. [8] reported the isolation and structure of five microcystins from a strain of *Microcystis aeruginosa* that had been isolated from a hepatotoxic water bloom sample collected in Lake Kroshnoseso, Karelia, Russia.

This paper presents the results of a toxicity investigation on water bloom samples collected in Dnieper River basin reservoirs and in some water bodies of the European part of Russia. Also included are results of toxicity testing for a *M. aeruginosa* strain isolated from a Dnieper River water bloom.

2 MATERIALS AND METHODS

Phytoplankton Sampling and Strains Purification

Bloom samples were collected during the period June to August in 1989-1991, using a no. 25 plankton net. Dominant species in all toxic blooms were identified according to the system of Gollerbach et al. [9]. Collected material was either air dried in the field or lyophilized in laboratory and stored for use in toxicity testing and toxin extraction.

To obtain viable cyanobacterial strains small
quantities of the water bloom were spread into glass dishes
containing solid Z8 media [10] directly in the field. In the
laboratory, colonies from the dishes were transferred to
Eppendorf tubes containing Z8 media. Viable cultures from
the isolates were chosen for further investigation.

Acute Toxicity Bioassay

Toxicity tests were carried out using intraperitoneal
(i.p.) injection of white male mice (15-30 g) with
lyophilized or air dried cells suspended in distilled water.
Signs of poisoning, survival times, body weights and liver
weights were recorded. Mouse tests were also used for
toxicity testing of fractions collected during purification
of the toxic cells.

Toxin Purification

Isolation and purification of toxins from bloom samples
and laboratory cultures were done according to a modified
HPLC method as described by Krishnamurthy et al. [11].

Toxin Analysis

Toxins purified by HPLC were analyzed using UV
absorption spectroscopy, amino acid composition [11] and fast
atom bombardment mass spectrometry (FAB-MS) [11, 12].

3 RESULTS

All toxic cyanobacteria samples used for toxin isolation are
listed in Table 1.

4 DISCUSSION

Our results expand the existing body of information which
demonstrates that toxic cyanobacteria are present in water
bodies of the former Soviet Union. To date only the potent
hepatotoxic tumor promoting microcystins have been identified
from the various toxic waterbloom samples and laboratory
isolates studied. Toxicity of all the *M. aeruginosa* samples
tested in the mouse bioassay show characteristic signs of
poisoning that are the same as that reported for the
microcystins [13]. Histopathology examination of the mouse
livers show disaggregated hepatocytes and other changes
specific to the microcystins [14].

Mouse bioassay and enzyme linked immunosorbent assay
(ELISA) - guided HPLC separation of the toxic fraction
revealed the presence of from 1-6 microcystins in the various
samples and isolated strains. *M. aeruginosa* strain MAK-5
from the Kiev city reservoir contains at least six
microcystins, two which are yet to be identified. In
contrast the two waterbloom samples from Lake Razliv contain
only a single microcystin (Table 1).

Table 1   Sample location, dominant species and toxin composition of peptide liver toxin from cyanobacteria collected in water bodies of Russia and The Ukraine

Sample number	Sampling site	Date of sampling	Dominant species	Toxin composition
1	River Dnieper, Kiev	08/08/89	*M. aeruginosa*	Two toxins - structure not established
2	Glebov Gulf, Kiev reservoir	08/10/89	*M. aeruginosa*	MCYST-YR, MCYST-LR -3 desmethyl, MCYST-LR -3 desmethyl MCYST-RR one toxin structure not established
3	Sea Club, Kiev reservoir	08/11/89	*M. aeruginosa*	MCYST-YR, MCYST-LR one toxin structure not established
4	Rasliv Lake, St. Petersburg	09/20/90	*M. aeruginosa*	One toxin structure not established
5	Rasliv Lake, St. Petersburg	07/20/91	*M. aeruginosa*	MCYST-LR one toxin structure not established
6	Kursh Gulf, Kaliningrad	08/08/91	*M. aeruginosa*	Two toxin structures not established
7	MAK-5, laboratory culture isolated from a bloom in the Kiev city drinking water reservoir		MCYST-RR, 7-desmethyl	MCYST-YR, MCYST-LR, MCYST-RR and two toxins with structure not established

In addition to the samples listed in Table 1 the
following samples and isolates are currently being examined.
1) Two hepatotoxic strains of *M. aeruginosa* from a bloom
sample collected in Ladoga Lake, northeast of St. Petersburg.
This lake serves as a drinking water supply for much of the
region including parts of St. Petersburg. The microcystins
from these strains are currently being investigated. 2)
Toxicity of waterbloom samples from the Gulf of Finland (St.
Petersburg recreation zone) and Lake Down Suzdal (St.
Petersburg recreation zone) both collected in July 1991. The
dominant species of cyanobacteria present was *Aphanizomenon
flos-aquae*. Toxicity testing by the intraperitoneal mouse
bioassay indicated only a low level of hepatotoxicity.

All water bodies investigated in our study are
important for fishing, transportation, recreation or as
drinking water supplies. All have become eutrophic to
hypereutrophic due to nutrient enrichment from human activity
(Kursh Gulf, Rasliv Lake) or due to artificial impoundments
and the resulting changes in their nutrient status (River
Dnieper reservoirs, Kiev) [15]. This nutrient enrichment has
led to annual blooms in the Kursh Gulf and Rasliv Lake so
severe that these water bodies have become almost useless for
recreation purposes.

Another human health danger emerges when these water
bodies containing toxic cyanobacteria are used as drinking
water supplies. There is an increasing body of experimental
data which shows that microcystins promote cancer in
laboratory test animals [16] and act as potent promoters of
liver tumors [17].

Available data on phytoplankton species [15, 18] indicate
that toxigenic cyanobacteria waterblooms are a regular
occurrence in water bodies of the European and Asiatic parts
of Russia. It is recommended that the use of analytical and
biological methods for the detection of these toxins can help
to establish regular monitoring of microcystins in affected
water bodies of Russia. Regulations concerning the
monitoring of water containing microcystins should be
developed. A national program in Russia to inform the public
about the possibility of cyanobacteria intoxication is also
recommended.

5. ACKNOWLEDGMENTS

Isolation and characterization of microcystins was supported
by a Wright State University travel grant to V.M.T. Mass
spectrometry analysis was provided in part by a grant from
the National Institute of General Medical Sciences to K.L.
Rinehart, University of Illinois and a subcontract from this
grant to W.W. Carmichael, Wright State University. V.M.T.
would like to thank the following persons for their support
and collaboration: W.W. Carmichael and W.R. Evans, Wright
State University, Dayton, Ohio; K.L. Rinehart and M.
Namikoshi, University of Illinois, Urbana, Illinois.

REFERENCES

1. G. Francis, <u>Nature</u> (London), 1878, <u>18</u>, 11.
2. O.M. Skulberg, G.A. Codd and W.W. Carmichael, <u>Ambio</u>, 1984, <u>13</u>, 244.
3. W.W. Carmichael, C.L.A. Jones, N.A. Mahmood and W.C. Theiss, <u>CRC Crit. Rev. Envir. Control</u>, 1985, <u>15</u>, 275.
4. W.W. Carmichael, <u>J. Appl. Bacteriol.</u>, 1992, <u>72</u>, 445.
5. W.W. Carmichael, N.A. Mahmood and E.G. Hyde, <u>Marine Toxins: Origin, Structure and Molecular Pharmacology</u>, Am. Chem. Soc. Symp. Series #<u>418</u>, 1990, 87.
6. Y.A. Kirpenko, I.I. Perevozchenko, L.A. Sirenko and L.F. Lukina, <u>Dopov. Acad. Nayk. Ukr. RSR Ser. B</u>, 1975, 359.
7. Y.A. Kirpenko and N.I. Kirpenko, <u>Hydrobiol. J.</u>, 1980, <u>16</u>, 53. (in Russian)
8. K. Sivonen, M. Namikoshi, W.R. Evans, B.V. Gromov, W.W. Carmichael and K.L. Rinehart, <u>Toxicon</u>, 1992, <u>30</u>, 1481.
9. M.M. Gollerbach, E.K. Kosinskaia and V.I. Poliansky, <u>Blue-green Algae. Part 2</u>, Sovetskaia Nayka, Moscow, 1953.
10. R. Rippka, 'Methods in Enzymology', Academic Press, 1988, <u>167</u>, 3.
11. T. Krishnamurthy, W.W. Carmichael and E.W. Sarver, <u>Toxicon</u>, 1985, <u>24</u>, 865.
12. K.L. Rinehart, K-I. Harada, M. Namikoshi, C. Chen, C. Harvis, M.H.G. Munro, J.W. Blunt, P.E. Mulligan, V.R. Beasley, A.M. Dahlem and W.W. Carmichael, <u>J. Am. Chem. Soc.</u>, 1988, <u>110</u>, 8557.
13. P.R. Gorham and W.W. Carmichael, 'Algae and Human Affairs', Cambridge Univ. Press, 1988, 404.
14. A.S. Dabholkar and W.W. Carmichael, <u>Toxicon</u>, 1987, <u>25</u>, 285.
15. L.A. Sirenko, 'Vegetation and Bacterial Population of Dnieper and its Reservoirs', Naukova Dumka, Kiev, 1989, 98 (in Russian).
16. I.R. Falconer, <u>Environ. Toxicol. Water Qual.</u>, 1991, <u>6</u>, 177.
17. R. Nishiwaki-Matsushima, T. Ohta, S. Nishiwaki, M. Suganuma, K. Kohyama, T. Ishikawa, W.W. Carmichael and H. Fujiki, <u>J. Cancer Res. Clin. Oncol.</u>, 1992, <u>118</u>, 420.
18. V.I. Ermolaev, 'Phytoplankton of the Water Bodies of Lake Sartlan Drainage

Structural Analysis of Cyanobacterial Toxins

Ken-ichi Harada,[1] Makoto Suzuki,[1] and Mariyo F. Watanabe[2]

[1]FACULTY OF PHARMACY, MEIJO UNIVERSITY, TEMPAKU, NAGOYA 468, JAPAN

[2]TOKYO METROPOLITAN RESEARCH LABORATORY OF PUBLIC HEALTH, SHINJUKU, TOKYO 160, JAPAN

1 INTRODUCTION

It is well known that cyanobacteria produce numerous secondary metabolites that are not used for primary metabolism.[1] There may be two aspects associated with the production of secondary metabolites by cyanobacteria. Recently, cyanobacteria have become one of the important sources for bioactive substances, and their importance has increased considerably. Since 1970's, many types of compounds have been isolated and they show characteristic biological activities, cytotoxic, immunosuppressive, antifungal, cardioactive and enzyme inhibitory. Particularly, cytotoxic compounds such as scytophycins[2] and tantazoles[3] have been paid much attention.

On the other hand several metabolites produced by cyanobacteria show characteristic acute toxicity. The toxins from freshwater and brackish cyanobacteria are classified into groups according to their toxicity. Microcystins, nodularin and cylindrospermopsin are known as the lethal hepatotoxins and the neurotoxins, anatoxin-a, anatoxin-a(s) and aphantoxins are produced by some species of cyanobacteria. Poisoning cases by these toxins involve sickness and death of livestock, pets and wildlife following ingestion of water containing toxic algae cells or the toxin released by the aging cells.[1]

These toxins are produced by species and strains of planktonic cyanobacterial genera such as *Anabaena*, *Microcystis* and *Oscillatoria* which are commonly observed world-wide in eutrophic lakes. In many cases water in such lakes is being used for drinking water. Under these circumstances, it is essential to chemically know the distribution of cyanobacterial toxins under any field conditions. For this purpose, the following studies based on chemistry should be achieved along with many studies in other areas: establishment of an analysis method for qualification and quantification, isolation and structure determination of a new toxin and investigation of stability of the toxins and their detoxification.

2 ISOMERIZATION OF MICROCYSTINS TO NON-TOXIC GEOMETRICAL ISOMER

Microcystins are potent hepatotoxins produced by *Microcystis aeruginosa*, *M. viridis*, *Nostoc* sp. *Oscillatoria agardhii* and *Anabaena flos-aquae* and they are structurally monocyclic heptapeptides as shown in Fig. 1. Over 40 microcystins have been isolated so far and they show hepatotoxicity ranging from 50 to 800 µg/kg of LD_{50}.[4] It is also found that microcystins inhibit protein phosphatases 1 and 2A in a manner similar to okadaic acid and have a tumor-promoting activity on rat liver.[5]

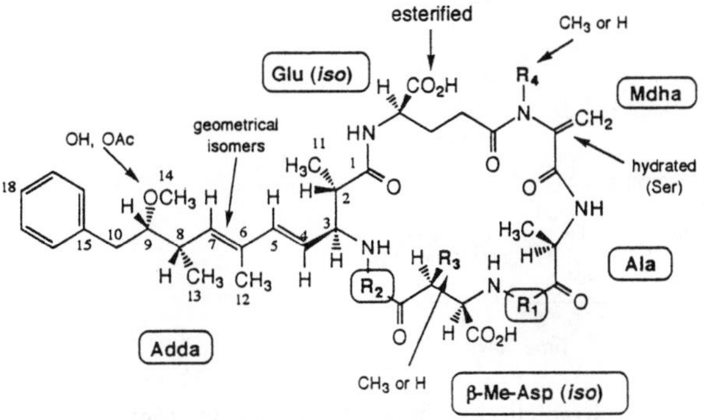

Fig. 1. Structural variations of microcystins.

Severe outbreaks of toxic cyanobacteria bloom have been observed in water supply reservoirs in many countries. Hughes *et al.*, reported that a toxic substance was detected in the culture filtrate during the early stage of the growth.[6] We have also observed the release of microcystins into the surrounding culture medium during the decomposition of *Microcystis aeruginosa*.[7] These findings suggest that microcystins are normally confined within the cyanobacterial cells and enter into surrounding water after lysis and cell death under field conditions. However, the amount of microcystins detected in lake water was at most a few μg/l, and the amount was much less than that estimated in cells. To assess the health implications, it is very important to pursue microcystins under field conditions. Five pathways may be considered to contribute to natural routes of detoxification of microcystins:

(1) dilution
(2) adsorption
(3) thermal decomposition aided by temperature and pH
(4) photolysis
(5) biological degradation

Sunlight irradiates the earth at wave lengths above 295 nm and is essential for growth of cyanobacteria. Although influence of fluorescent light and natural sunlight on stability of microcystin LR was observed in distilled water for 26 days, no significant change was found. Cyanobacteria possess several pigments for photosynthesis such as chlorophyll *a*, β-carotene and phycocyanins. If cells decompose under field conditions, microcystins would be exposed to sunlight together with coexisting pigments. Fig. 2 illustrates the decrease of microcystin by irradiation with sunlight for 15 days in the presence of various pigments, indicating that the presence of pigments accelerates the decomposition. However, no significant decomposition of microcystin LR occurred in pigment solutions by irradiation with fluorescent light.

Fig. 3 shows a typical HPLC chromatogram of the photolysis product of microcystin LR after 7 days and one new peak appears, together with the starting material. A co-elution experiment indicated clearly that this peak corresponds to a geometrical isomer, 4(*E*), 6(*Z*) isomer of the diene of Adda portion in microcystin LR (abbreviated as 6(*Z*)-Adda microcystin LR). We have isolated these geometrical isomers of microcystins LR and RR in the course of isolation of microcystins from natural blooms of *Microcystis* and their structures have been determined by extensive 2D NMR experiments.[8,9] Toxic bloom samples usually contain 5 to 15 % of the geometrical isomers in Japan. The isolated isomers do not show hepatotoxicity[8] and have much weaker tumor promoting activity than their parent toxins,[10] indicating that the 4(*E*), 6(*E*)-Adda portion is essential for these biological activities.

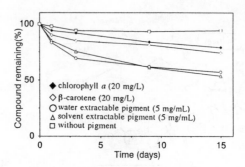

Fig. 2. Decrease of microcystin LR by irradiation with sunlight in various pigment
solutions.

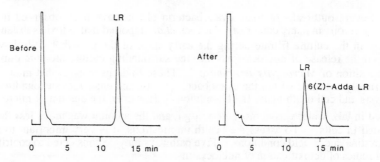

Fig. 3. HPLC chromatograms of microcystin LR before and after 7 days irradiation with
sunlight at 5 mg/mL of water extractable pigment.

The time course of the isomerization for microcystin LR and its isomer is shown
in Fig. 4. Both isomers were gradually isomerized to the corresponding ones and the
reactions reached an equilibrium after 18 days in the presence of water extractable
pigments.   The isomerization ratios in equilibrium were approximately 0.55.
Subsequently, the effect of the concentration of water extractable pigments on the
isomerization was observed for 29 days.   Although linear relationships between
pigment concentration and isomerization ratio were obtained after 6 and 8 days , the
relationship was not found over 8 days due to complete decomposition of both isomers.
These results indicated that the decomposition and isomerization of microcystin LR
occur simultaneously under these conditions and the former is predominant at higher
pigment concentrations.   The isomerization was influenced by the presence of pigment
in water and its rate was dependent on concentration and the type of pigment.[11]

The photolysis product of microcystin LR by irradiation with sunlight in the
presence of water extractable pigment was analyzed using the Frit-FAB LC/MS method.
As shown in Fig. 5 three peaks, a major peak A and two minor peaks, X and Y, are
present in addition to that of microcystin LR in the total ion chromatogram. Peak A can
be easily identified to be 6(Z)-Adda microcystin LR by its retention time, mass
chromatography monitored at m/z 995 and mass spectrum.   Mass chromatography at
m/z 135 is a very powerful technique for identification of microcystins. Peaks X and Y
are detected by this technique, and their molecular weights are 1028, suggesting that they
have two hydroxyl groups oxidatively added in the diene group of Adda.   They can be
key compounds in the elucidation of a mechanism for the isolation of microcystins by
photolysis with pigment.

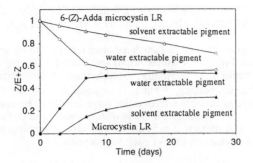

Fig. 4. Isomerization of microcystin LR and 6(Z)-Adda microcystin LR by irradiation
    with sunlight in various pigment solutions.

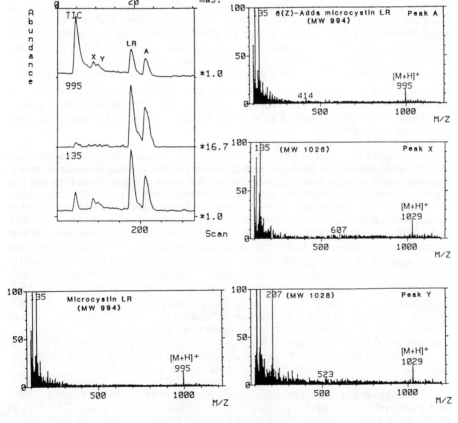

Fig. 5. Frit-FAB LC/MS analysis of photolysis products of microcystin LR.

3 EFFECTIVE USE OF FRIT-FAB LC/MS FOR SCREENING AND IDENTIFICATION OF MICROCYSTIN AND NODULARIN

As mentioned above, over 40 microcystins have been reported so far and their structural variations are shown in Fig. 1.   A simple, rapid and precise screening and identification technique is highly required.   Originally, the structures of microcystins were mainly determined by NMR spectroscopy,[12] but it is impossible to apply the technique for identification of the toxins because microcystins have always to be purified and it needs several mg of pure toxins.   Although HPLC is widely used for detection and separation, it cannot always lead to a definite conclusion in the case of treating many components because it relies on only retention time of each microcystin.

Fast atom bombardment mass spectrometry (FABMS) and liquid secondary ion mass spectrometry (LSIMS) are powerful techniques for obtaining molecular weight information on polar and involatile compounds.   Standard FAB and LSIMS have been used for the determination of the molecular weights of microcystins and nodularin. However, it is difficult to obtain sequence information about constituent amino acids because microcystins and nodularin are cyclic peptides.   Additionally, there appears to be considerably greater background ions, which are originating from matrices, interfering with the characterization of structurally informative fragment ions in the lower mass region.   We recently described a new analytical method for microcystins using Frit-FAB LC/MS, which allowed the rapid identification of microcystins and related compounds.[13]   The method has also provided another advantage that the background ions can be subtracted to give the mass spectra consisting of sample ions only.   We have found that the characteristic fragment ion at m/z 135, which is formed by α-cleavage of a methoxy group of Adda, was observed with considerable abundance in the background subtracted mass spectra of microcystins.   The mass chromatography monitored at m/z 135 has proved to be useful for differentiation of microcystins from other types of compounds.

As shown in the previous section the LC/MS technique was applied successfully to characterize photolysis products of microcystin LR and is being used for structural characterization of microcystins in complicated bloom samples and reaction products in various stability tests of microcystins.   Fig. 6 shows the Frit-FAB LC/MS analysis data of a toxic fraction from bloom sample collected in South Australia.   The toxic fraction contains several microcystins as shown by the mass chromatogram at m/z 135.   Peaks 1 and 2 were easily determined to be microcystins RR and LR, respectively, according to the mass spectra and mass chromatograms at their $[M+H]^+$.   The mass spectrum of peak 3 shows the $[M+H]^+$ at m/z 1068 corresponding to microcystin WR.   It is known that immonium ions originating from constituent amino acids of peptides frequently appear in the lower mass region of FAB mass spectra.[14]   This is another advantage that immonium type ions can be detected in the background subtracted spectra by Frit-FAB LC/MS.   In the case of peak 3 the spectrum indicates the immonium ions at m/z 112 and 70 from arginine and m/z 159 from tryptophan, affording the further confirmation. Although the $[M+H]^+$ ion at m/z 1029 for peak 4 corresponds to microcystin M(O)R or FR, it could not be conclusively identified, because the immonium ions of methionine sulfoxide and phenylalanine have the same mass value at m/z 120.

An application of the technique made it possible to isolate types of peptides other than microcystins from the toxic fractions of two *Microcystis aeruginosa* TAC 95 and M228; these mainly produce microcystins LR and YR, respectively.   The structures of four cyclic depsipeptides, aeruginopeptins (**1-4**), are shown in Fig. 7, which is difficult to detect by usual HPLC with UV (238 nm).[15]

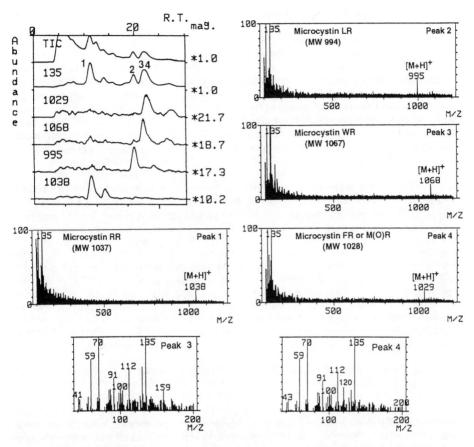

Fig. 6. Frit-FAB LC/MS analysis of a toxic fraction from a bloom sample collected in South Australia.

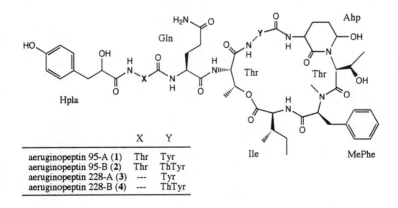

	X	Y
aeruginopeptin 95-A (**1**)	Thr	Tyr
aeruginopeptin 95-B (**2**)	Thr	ThTyr
aeruginopeptin 228-A (**3**)	---	Tyr
aeruginopeptin 228-B (**4**)	---	ThTyr

Fig. 7. Structures of aeruginopeptins.

4 LC/MS DETECTION OF ANATOXIN-A

Anatoxin-a is the first toxin obtained from a freshwater cyanobacterium, *Anabaena flos-aquae* to be chemically defined as the secondary amine, 2-acetyl-9-azabicyclo[4.2.1]non-2-ene with the molecular weight of 165. It is a potent nicotinic agonist which acts as a postsynaptic, depolarizing and neuromuscular blocking agent, with high toxicity.[1] Anatoxin-a has occurred occasionally in North America and northern Europe and no occurrence has been reported in Japan so far. The toxin is relatively unstable especially under basic conditions, and Stevens and Krieger have studied the stability under field conditions.[16]   Additionally, it is known that *Anabaena flos-aquae* produces simultaneously a neurotoxin and hepatotoxins.[17]   So a suitable analysis method has been required for the investigation of distribution and detoxification of anatoxin-a.

There have been several analysis methods for anatoxin-a and we have also established an analysis system including solid-phase extraction and HPLC with UV detection.[18]   As an example, Fig. 8 shows the HPLC chromatogram/UV (227 nm) of a toxic fraction of a culture strain by our method. It is probably, however, difficult to accurately quantify the amount of anatoxin-a indicated by the arrow, because the amount contained in the toxic fraction is very limited and the fraction includes many contaminants. To analyze such samples a more sensitive and specific method has been required.   As shown in the previous section a method combining HPLC and mass spectrometry using an appropriate interface would also offer significant advantages in this case.   Thermospray (TSP) is a relatively older interface than recently developed ones such as electrospray, but has been widely used because whole effluent can be introduced into the mass spectrometer.   TSP-LC/MS would be expected to provide a sensitive quantification and specific detection of anatoxin-a in various samples.

It is frequently pointed out that it is difficult relatively to obtain reproducible results by TSP-LC/MS.   This may be mainly caused by its vacuum system and heating problems of the interface.   To overcome these problems and to obtain an accurate analysis result, an internal standard, acetyltropin, was introduced. As the result of extensive optimization of TSP and HPLC conditions, the appropriate operating conditions were established.   A combination of the optimized operating conditions and our clean-up method including a solid phase extraction with a reversed phase carboxylic acid cartridge made possible a sensitive, specific and reproducible analysis of anatoxin-a and its non-toxic oxidation product (MW 181). In addition, the LC/MS provided an excellent linearity between the concentration of anatoxin-a and peak heights.[19]

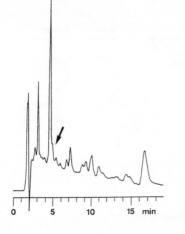

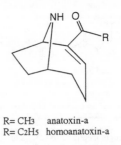

R= CH3   anatoxin-a
R= C2H5   homoanatoxin-a

Fig. 8. HPLC chromatogram of a toxic fraction of TAC210 (arrow indicates anatoxin-a).

Fig. 9 shows the TSP-LS/MS analysis data monitored at m/z 166 for anatoxin-a, m/z 182 for its epoxide and m/z 184 for the internal standard. In comparison with the HPLC chromatogram in Fig. 8, anatoxin-a was clearly detected by this method, along with two other compounds. Subsequently, the established method was applied to analysis of anatoxin-a in various foreign and Japanese samples. As mentioned earlier, anatoxin-a has not been discovered so far in Japan. Three strains and two bloom samples from Japanese lakes were investigated by the method and we could detect trace amounts of anatoxin-a in these samples. Recently, Skulberg *et al.*, reported the isolation and structure determination of homoanatoxin-a from *Oscillatoria formosa*.[20] Although the TSP-LC/MS was applicable to a screening of the toxin without the standard sample, no toxin was detected by SIM of m/z 180 in all samples of the present study.

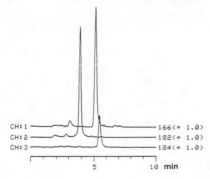

Fig. 9. Selected ion monitoring of anatoxin-a and its epoxide from TAC210, and internal standard at m/z 166, 182 and 184, respectively, under TSP-LC/MS conditions.

5 ISOLATION OF CYLINDROSPERMOPSIN FROM A CYANOBACTERIUM *UMEZAKIA NATANS*

In 1987 *Umezakia natans* (Stigonemataceae) was isolated from net samples collected at Lake Mikata, Fukui, Japan.[21] It showed the following characteristics: i) T-shaped branching instead of reverse V-shaped branching, ii) gelatinous sheath instead of firm sheath, iii) latent faculties of producing many spores and heterocysts. We succeeded in mass cultivation of this organism and found that the alga shows hepatotoxicity to mice. *U. natans* was cultured in our laboratory and was harvested by plankton net sieve. The methanol extract of lyophilized alga exhibited toxicity to mice at 1200 mg/kg. The extract was fractionated by successive reversed-phase chromatography, HP-20 column chromatography, and reversed-phase HPLC to furnish a toxin in 0.09 % yield. The toxin showed toxicity to mice at 7 mg/kg that accounts for all the toxicity.

The positive-ion FAB mass spectrum with glycerol matrix displayed the [M+H]$^+$ at m/z 416, and the negative-ion FAB mass spectrum with glycerol matrix displayed the [M-H]$^-$ at m/z 414, indicating the molecular weight of 415 daltons for the toxin. In the ^{1}H NMR and ^{13}C NMR spectra, the following protons and carbon signals were observed, respectively: ^{1}H NMR (D$_2$O) δ: 0.97 (3H, d, J=6.6 Hz), 1.50 (1H, brt, J=14 Hz), 1.55 (1H, brq, J=12.5 Hz), 1.83 (1H, m), 2.14 (1H, dt, J=13.4, 4 Hz), 2.43 (1H, dt, J=14.5, 4 Hz), 3.23 (1H, brt, J=10 Hz), 3.64 (1H, m), 3.72 (1H, m), 3.83 (1H, brt, J=9 Hz), 3.85 (1H, m), 4.58 (1H, m), 4.67 (1H, m), 5.81 (1H, s), ^{13}C NMR (D2O) δ: 13.4, 28.1, 35.9, 39.4, 44.7, 48.0, 53.3, 57.5, 70.0, 77.8, 99.6, 153.6, 156.2, 167.5.

    The UV spectrum (λ_{max} 262 nm in water) and a singlet ^{1}H signal at δ 5.81 strongly suggested the presence of a substituted uracil moiety. The detailed analysis of the ^{1}H-^{1}H COSY (correlation spectroscopy) spectrum led to partial structures shown in Fig. 10. The *B/E* linked scan of the [M+H]+ at *m/z* 416 under FABMS conditions gave three informative ions, *m/z* 336 [M+H-SO₃]+ , *m/z* 318 [M+H-H₂SO₄]+, and *m/z* 274 [M+H-C₅H₆N₂O₃]+ (Fig. 11), indicating that the toxin possesses a sulphate ester and hydroxymethyluracil [C₅H₅N₂O₃] moieties. These spectral data are quite similar to those reported for cylindrospermopsin which is produced by *Cylindrospermopsis raciborskii* in Australia and was responsible for human hepatoenteritis.[22] Detailed comparison of the NMR data of the present toxin and those of cylindrospermopsin[23] showed that they are completely identical. Thus, the toxic component from *U. natans* turned out to be cylindrospermopsin. This is the second case of isolating cylindrospermopsin from cyanobacteria, but the first one in Japan.[24] The detailed study on pathological and toxicological aspects by cylindrospermopsin is in progress.

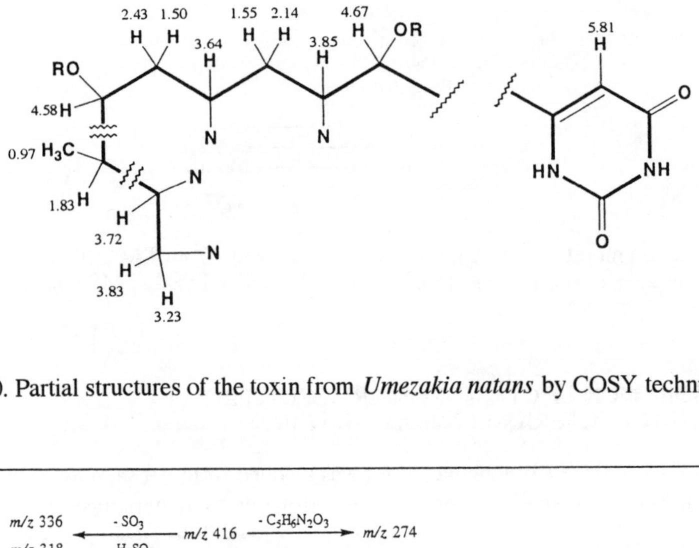

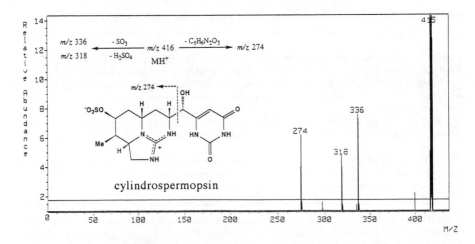

Fig. 10. Partial structures of the toxin from *Umezakia natans* by COSY technique.

Fig. 11. Product ion spectrum of m/z 416 [M+H]+ of the toxin from *Umezakia natans* under FABMS conditions.

REFERENCES

1.   W.W. Carmichael, *J. Appl. Bact.*, 1992, **72**, 445.
2.   M. Ishibashi, R.E. Moore and G.M.L. Patterson, *J. Org. Chem.*, 1986, **51**, 5300.
3.   S. Carmeli, R.E. Moore and G.M.L. Patterson, *J. Am. Chem. Soc.*, 1990, **112**, 8195.
4.   R.R. Stotts, M. Namikoshi, W.M. Haschek, K.L. Rinehart, W.W. Carmichael, A.M. Dahlem and V.R. Beasley, *Toxicon*, 1993, **31**, 783.
5.   R. Nishiwaki-Matsushima, T. Ohta, S. Nishiwaki, M. Suganuma, K. Kohyama, T. Ishikawa, W.W. Carmichael and H. Fujiki, L. *Cancer Res. Cli. Oncol.*, 1992, **118**, 420.
6.   E.O. Hughes, P.R. Gorham and A. Zender, *Can. J. Microbiol.*, 1958, **4**, 225.
7.   M.F. Watanabe, K. Tsuji, Y. Watanabe, K.-I. Harada and M. Suzuki, *Natural Toxins*, 1992, **1**, 48.
8.   K.-I. Harada, K. Matsuura, M. Suzuki, M.F. Watanabe, S. Oishi, A.M. Dahlem, V.R. Beasley and W.W. Carmichael, *Toxicon*, 1990, **28**, 55-64.
9.   K.-I. Harada, K. Ogawa, K. Matuura, H. Murata, M. Suzuki, M.F. Watanabe, Y. Itezono and N. Nakayama, *Chem. Res. Toxicol.*, 1990, **3**, 473.
10. R. Nishiwaki-Matsushima, S. Nishiwaki, T. Ohta, S. Yoshizawa, M. Suganuma, K-I. Harada, M.F. Watanabe and H. Fujiki, *Jpn. J. Cancer Res.*, 1991, **82**, 993.
11. K. Tsuji, S. Naito, F. Kondo, N. Ishikawa, M.F. Watanabe, M. Suzuki and K.-I. Harada, *Environ. Sci. Tech.*, 1994, **28**, 173.
12. D.P. Botes, A.A. Tuinman, P.L. Wessels, C. C Viljoen, H. Kruger, D.H. Williams, S. Santikarn, R. J. Smith and S.J. Hammond, *J. Chem. Soc.* Perkin Trans. I, 1984, 2311.
13. F. Kondo, Y. Ikai, H. Oka, N. Ishikawa, M.F. Watanabe, M. Watanabe, K.-I. Harada and M. Suzuki, *Toxicon*, 1992, **30**, 227.
14. K. Sato, T. Asada, M. Ishihara, F. Kunihiro, Y. Kammei, E. Kubota, C.E. Costello, S.A. Martin, H.A. Scoble and K. Biemann, *Anal. Chem.*, 1987, **59**, 1652.
15. K.-I. Harada, T. Mayumi, T. Shimada, M. Suzuki, F. Kondo and M.F. Watanabe, *Tetrahedron Lett.*, 1993, **34**, 6091.
16. D.K. Stevens and R.I. Krieger, *Toxicon*, 1991, **29**, 167.
17. K.-I. Harada, K. Ogawa, Y. Kimura, H. Murata, M. Suzuki, P.M. Thorn, W.R. Evans and W.W. Carmichael, *Chem. Res. Toxicol.* 1991, **4**, 535.
18. K.-I. Harada, Y. Kimura, K. Ogawa, M. Suzuki, A.M. Dahlem, V.R. Beasley and W.W. Carmicheal, *Toxicon*, 1989, **27**, 227.
19. K.-I. Harada, H. Nagai, Y. Kimura, M. Suzuki, H. Park, M.F. Watanabe, R. Luukkainen, K. Sivonen and W.W. Carmichael, *Tetrahedron*, 1993, **41**, 9251.
20. O.M. Skulberg, W.W. Carmichael, R.A. Andersen, S. Matsunaga, R.E. Moore and R. Skulberg, *Environ. Toxicol. Chem.*, 1992, **11**, 321.
21. M. Watanabe, *Bull. Natn. Sci. Mus., Tokyo Ser.*, 1987, **B13**, 81.
22. P.R. Hawkins, M.T.C. Runnegar, A.R.B. Jackson and I.R. Falconer, *Appl. Environ. Microbiol.*, 1985, **50**, 1292.
23. I. Ohtani, R.E. Moore and M.T.C. Runnegar, *J. Am. Chem. Soc.*, 1992, **114**, 7941.
24. K.-I. Harada, I. Ohtani, K. Iwamoto, M. Suzuki, M.F. Watanabe, M. Watanabe and K. Terao, *Toxicon*, 1994, **32**, 73.

ACKNOWLEDGEMENTS

We would like to thank the following persons for devoted help: H. Murata, I. Ohtani, K. Ogawa, H. Nagai, T. Mayumi, T. Shimada, K. Iwamoto (Meijo University), Y. Ikai, F. Kond, H. Oka (Aichi Prefectural Institute of Public Health), K. Tsuji (Kanagawa, Prefectural Public Health Laboratories), K. Terao (Chiba University), M. Watanabe (National Science Museum), H. Park (Shinshu University), K. Sivonen (University of Helsinki) and W. W. Carmichael (Wright State University).

Determination of Anatoxin-a, Homoanatoxin and Propylanatoxin in Cyanobacterial Extracts by HPLC, GC–Mass Spectrometry and Capillary Electrophoresis

Terry M. Jefferies,[1]* Gavin Brammer,[1] Anastasia Zotou,[1,3]
Paul A. Brough,[2] and Timothy Gallagher[2]

[1]SCHOOL OF PHARMACY & PHARMACOLOGY, UNIVERSITY OF BATH,
BATH BA2 7AY, UK

[2]SCHOOL OF CHEMISTRY, UNIVERSITY OF BRISTOL, BRISTOL BS8 1TS, UK

[3]PRESENT ADDRESS: LABORATORY OF ANALYTICAL CHEMISTRY, CHEMISTRY
DEPARTMENT, UNIVERSITY OF THESSALONIKI, 54006 THESSALONIKI, GREECE

1 INTRODUCTION

Anatoxin-a (AnTx), 2-acetyl-9-azabicyclo[4.2.1]non-2-ene, is a bicyclic secondary amine, relative molecular mass 165, pK_a 9.4, incorporating an α,β-unsaturated enone moiety.[1] Most reports of AnTx occurrence are associated with *Anabaena flos-aquae, A.spiroides* or *A.circinalis.*[2] Recently, it has been shown that *Oscillatoria* strains also produce Antx.[3,4] Deaths of four dogs occurred during 1990 and 1991 due to cyanobacterial poisoning at Loch Insh, near Kingussie in the Grampian region of Scotland. *Oscillatoria* species were found at the water's edge and AnTx was identified in the bloom material and in the stomach contents of two of the poisoned dogs.[3] Homoanatoxin (HomoAnTx) is a methylene homologue of AnTx, produced by *Oscillatoria* species[5] having a similar, but less toxic, neuromuscular blocking activity to AnTx.[2,6]

Figure 1 (a)Anatoxin-a,(b)Homoanatoxin,(c)Propylanatoxin.

AnTx is not readily isolated in reasonable amounts from *Anabaena*, which is also not a reliable source of supply. HomoAnTx has only recently been identified in *Oscillatoria*[5] and is not commercially available. Propylanatoxin (PrAnTx) has not been reported as a naturally occurring toxin. AnTx and HomoAnTx occur naturally as the (+)-enantiomers, but as the analytical methods employed in this study do not distinguish between the (+) and (−) enantiomers, the (±)-racemic standards are equally acceptable. The compounds (±)-AnTx, (±)-HomoAnTx and (±)-PrAnTx, Fig. 1, were synthesised on the milligram scale as their hydrochloride salts at the University of Bristol by methods published elsewhere[6] and their high purity confirmed by RP-HPLC of aqueous solutions. This paper describes some chromatographic methods employed by the Authors for the analysis of these compounds in the presence of algal extracts.

2 EXPERIMENTAL

HPLC and GC-MS equipment, reagents and procedures were as reference 7. CE equipment was kindly loaned by Dionex UK Ltd, Leeds, UK.

3 RESULTS AND DISCUSSION

The properties of AnTx, HomoAnTx and PrAnTx that are important chromatographically are that they are highly water-soluble organic amines, stable in acidic aqueous solutions with a UV λmax about 227 nm. They are thus suitable for reversed-phase HPLC conditions, but the percentages of acetonitrile typically used (30-60 %)for the analysis of extracts containing microcystin-LR cause AnTx to be eluted with the solvent front. AnTx requires a low level of organic modifier in order to achieve reasonable retention. For example, by using a μBondapak C18 column (300 x 3.9 mm i.d), with a mobile phase of acetonitrile-0.05% trifluoroacetic acid (10 + 90 v/v), k' values of 1.2 and 3.4 were obtained for AnTx and HomoAnTx respectively. However under these conditions, a simple aqueous extract of an algal bloom, or a concentrated extract (200 μl) prepared from a 250 ml reservoir water sample following solid phase extraction of, for example, microcystin-LR, will produce a large number of unidentifiable peaks eluting with similar retentions to AnTx and HomoAnTx.

Retention can be more selectively increased by the addition to the mobile phase of an ion-pair reagent such as sodium dodecylsulphate, which is distributed to the hydrocarbon surface of the packing material, making the surface negatively charged. Under acidic conditions, AnTx and its homologues are positively charged and form unstable complexes with the reagent in the mobile phase and on the hydrocarbon surface, and so retention is enhanced. For example, using a Hypersil-BDS column with a mobile phase containing 30% acetonitrile, k' values of 7.3, 11.7, and 20.3 were obtained for AnTx, HomoAnTx and PrAnTx, respectively, Fig 2a. This provides better conditions for these compounds to be eluted free from interferences, although the extent of the interferences is highly variable, depending upon the sample and the processing method. For example, a *Microcystis* bloom was found to produce more potential interferences than an *Oscillatoria* bloom using identical sample processing and HPLC conditions. AnTx was readily identified in the latter bloom, and it was estimated

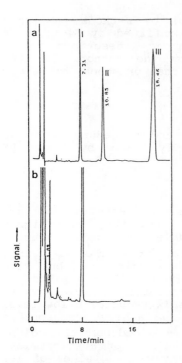

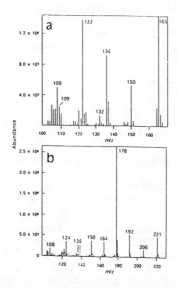

Figure 3 GC-MS of (a)Anatoxin-a
(b)N-butylanatoxin, from ref 7.

Figure 2 HPLC of (a) standard solutions of Anatoxin-a (I),
6μg/ml; Homoanatoxin (II), 9μg/ml; and Propylanatoxin (III)
about 12μg/ml. (b) 5mg freeze-dried benthic *Oscillatoria*
species (Loch Insh, 1991, G.A.Codd) extracted with 300μl
mobile phase by sonication for 5 mins, centrifuged for 10
mins, and 10μl injected. Conditions: Hypersil-BDS 150 x
4.6mm id column at 30°C with acetonitrile-0.05mM SDS in
0.005M phospate buffer (pH 3)(30:70 v/v). Flow rate
1ml/min, UV at 227nm, AT 128.

that the freeze-dried material contained approximately 2 mg
AnTx per gram, Fig 2b. No evidence for the presence of
HomoAnTx was seen. Using this column, linear calibrations
were obtained using peak heights for AnTx, (0.07 to 5.4 μ
g/ml) and for HomoAnTx (0.18 to 7 μg/ml) with correlation
coefficients > 0.999. The selection of a base-deactivated
column material (5 μm) also improves the chromatography,
producing narrow peaks and high column efficiencies.

The GC-MS of anatoxin-a can be achieved without
derivatisation by selection of a cool injection temperature
(100°C) into a non-polar capillary column (e.g. HP-1, 50 m x
0.25 mm i.d,) temperature programmed from 60 to 240°C at

20°C/min.[7]   The retention time was 8.6 min and principal peaks occurred at m/z 165 (M+), 122, 136, 150, 108, 105, 109 and 132, Fig.3a.   However, it is preferable to derivatise amines due to the acidic nature of the silica capillary column, and the N-butyl derivative was chosen, rather than the acetyl derivative, in order to increase GC and HPLC retentions further.   Under the same GC conditions as before, retention increased to 10.2 min with principal peaks at m/z 178, 221 (M+), 192, 164, 150, 136, and 124, Fig 3b.   The series m/z 221, 206, 192, 178 and 164 shows the progressive loss of methylene units from the N-butyl group.   The enhanced stability of the derivatised AnTx compared to the underivatised AnTx can be assessed by the abundance of a single ion (m/z 178) Fig. 3b compared to the extensive fragmentation in Fig.3a.   Using this approach, AnTx was isolated by RP-HPLC from about 5 mg freeze-dried *Oscillatoria* bloom, the N-butyl derivative formed, and GC-MS carried out in Selected Ion Mode (SIM) to enhance sensitivity.   N-butylAnTx was identified by the presence of m/z ions 221, 164, 122, 136 and 150, all with identical retention times of 10.2 min.[7]

The technique of capillary electrophoresis (CE) is complimentary to HPLC and GC because the retention mechanism is completely different.   Compounds are separated by differences in their electrophoretic mobilities, which reflect differences in molecular size and charge.   The use of short (65 cm), small i.d. (0.075 mm) silica columns reduces the heat generated when high voltages (20 kV) are applied.   Uncoated silica columns are negatively charged above pH about 4 and attract the positive ions present in buffer solutions.   When the voltage is applied, these hydrated positive ions migrate to the negative electrode, transporting all the liquid in the column, including neutral sample compounds (electroosmotic effect).   Charged sample compounds also migrate to the negative or positive electrodes due to electrophoresis, so that they are eluted either before or after neutral compounds, repectively. Column efficiences are typically similar to those obtained in capillary GC.   Fig.4a shows that although AnTx, HomoAnTx and PrAnTx differ only by the presence of additional methylene units, these homologues can be baseline resolved using appropriate conditions.   AnTx was also readily identified at 4.8 min in a simple aqueous extract of *Oscillatoria*, Fig 4b, with unidentified algal peaks eluting mainly between 10-12 mins.   Peak heights of AnTx standard solutions were linear over the range 1.6 to 8 µg/ml (n=5), with linear regression of y = 4.3x - 0.5, r =0.9986.   By an approximate 1 in 5 dilution of the extract it was estimated that the freeze-dried material contained 4 mg AnTx/g.   The same CE conditions were also able to resolve AnTx and HomoAnTx from microcystin-LR and nodularin when added to an extracted reservoir water sample, Fig. 4c.   This is because the neurotoxins are small, positively charged compounds,

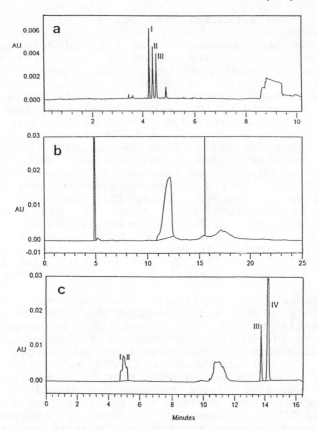

Figure 4 CE of (a) Anatoxin-a(I),homoanatoxin (II), and propylanatoxin (III), about 3µg/ml of each compound in 0.5mM HCl. (b) *Oscillatoria* extract prepared as Fig.2a (c) reservoir water sample (250 ml, not spiked) processed by solid phase extraction (amino and carboxylic acid cartridges) and reduced to 200µl. 100µl was spiked with anatoxin-a (I) (5µg/ml), homoanatoxin (II) (5µg/ml), microcystin-LR (III) (10µg/ml) and nodularin (IV) (10µg/ml). Conditions: Silica column 65cm x 75µm id containing 0.025M phosphate buffer (pH 5.4). Gravity injection, 100mm for 90sec. Constant voltage, 17kV. UV detection at 227nm.

whilst the endogenous peaks in the reservoir sample are predominately neutral (11-14 mins) and the hepatotoxins (13-15 mins) have a net negative charge. Nodularin is smaller than microcystin-LR and so is the last to be eluted. Preliminary quantitative studies have obtained linear calibrations for microcystin-LR and nodularin between 5 and

25 μg/ml, with r >0.99, and detection limits about 1 μg/ml and 0.25 μg/ml, respectively. The development of a single analytical CE method for neurotoxins and hepatotoxins is continuing.

4 CONCLUSIONS

HPLC with UV detection, GC-MS and CE with UV detection are all able to sensitively detect and measure neurotoxins in the presence of extracted algal components. HPLC is more sensitive than CE, but the neurotoxins are more readily resolved from algal components by CE. GC-MS remains the method of choice for peak identification.

REFERENCES

1. A.M.P.Koskinen and H.Rapoport, J.Med.Chem., 1985, 28, 1301.
2. W.W.Carmichael, J.Appl.Bacteriol., 1992, 72 445.
3. C.Edwards, K.A.Beattie, C.M.Scrimgeour and G.A.Codd, Toxicon, 1992, 30, 1165.
4. K.Sivonen, K.Himberg, R.Luukkainen, S.Niemela, G.K.Poon and G.A.Codd, Toxic. Assess., 1989, 4, 339.
5. O.M.Skulberg, W.W.Carmichael, R.A.Andersen, S.Matsunaga, R.E.Moore and R.Skulberg, Environ. Toxicol. Chem., 1992, 11, 321.
6. S.Wonnacott, K.L.Swanson, E.X.Albuquerque, N.J.S.Huby, P.Thompson and T.Gallagher, Biochem. Pharmacol., 1992, 43, 419.
7. A.Zotou, T.M.Jefferies, P.A.Brough and T.Gallagher, Analyst, 1993, 118, 753.

ACKNOWLEDGEMENTS

The financial support of Wessex Water Services Ltd. and Thames Water Utilities Plc for this project is gratefully acknowledged. The technical support of Kevin Smith, School of Pharmacy & Pharmacology, University of Bath, is gratefully acknowledged.

* To whom correspondence should be addressed.

Enantiomer-specific Analysis of Homoanatoxin-a, a Cyanophyte Neurotoxin

John-Erik Haugen,[1] Michael Oehme,[1] and Markus D. Müller[2]

[1]NORWEGIAN INSTITUTE FOR AIR RESEARCH, PO BOX 100, N-2007 KJELLER, NORWAY

[2]SWISS FEDERAL RESEARCH STATION, CH-8820 WÄDENSWIL, SWITZERLAND

1 INTRODUCTION

The blue-green algae *Oscillatoria* and *Anabaena* are among the most distributed toxin producing cyanophytes in eutrophicated freshwater. Consequently, toxic blooms of these species may have severe impact on water supply, fish farming, livestock and human health. Death of livestock, wildlife and pets caused by ingestion of water during blooms has been reported.[1-3] The neurotoxin produced by the *Oscillatoria* NIVA-CYA 92, named homoanatoxin-a due to its similarity to the toxin produced by *Anabaena*, has recently been isolated, its structure elucidated and its toxicity investigated.[4] Homoanatoxin-a is a low molecular weight bicyclic secondary amine. The exact structure of the toxin is 2-(propan-1-oxo-1-yl)-9-azabicyclo[4.2.1]non-2-ene (Figure 1). It has potent cholinergic properties and high toxicity. [4]

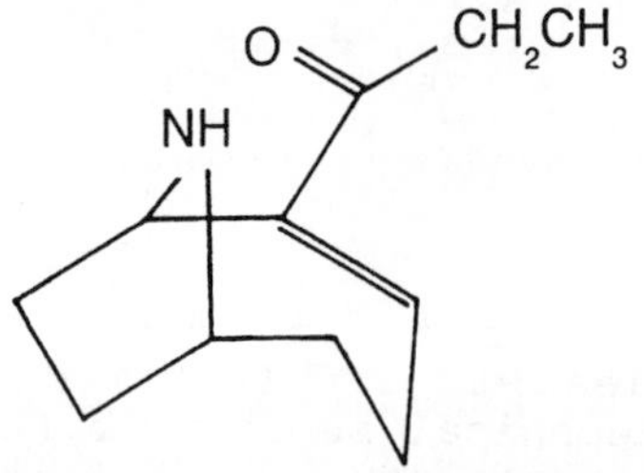

Figure 1 Structure of homoanatoxin.

Toxins produced by cyanophytes are often chiral and can exist as two optically active forms also called enantiomers. Enantiomers of a chiral compound have identical chemical and physical properties. They can only be transferred into each other by reflection. The ring structure of homoanatoxin contains two asymmetric centres which normally would result in two pairs of enantiomers. However, due to steric constraints of bond angles homoanatoxin exists in only two enantiomeric forms.[5] The chemical synthesis of such substances gives a racemate (a precise 1:1 mixture of both enantiomers) while biosynthesis is normally enantioselective producing only one enantiomer. Furthermore, in many cases one enantiomer has toxic properties while the other one is inactive or might even have antagonistic effects.

Separation of such chiral compounds into their enantiomers by chromatographic separation techniques is of great interest due to the following reasons:

(i) Confirmation of enantiomer purity of neurotoxins synthesized as racemates and separated into single enantiomers after formation of diastereomers by classical techniques.
(ii) Confirmation of enantiomer purity after chiral synthesis of neurotoxin.
(iii) Evaluation of the enantiomer-specificity of the biosynthesis by cyanophytes.

Recently, routine gas chromatographic methods have been developed which allow separation of enantiomers on special tailor-made stationary phases. Such phases consist of a chiral modificator such as modified cyclodextrins dissolved in a methyl-phenyl-polysiloxane.[6]

The aim of this study was to find a suitable chiral GC stationary phase for the separation of alkaloid neurotoxins and to elucidate the enantiomer specific biosynthesis of the toxin-producing cyanophyte *Oscillatoria*.

2 EXPERIMENTAL

The material investigated originated from a toxin producing blue-green alga *Oscillatoria formosa* strain NIVA-CYA 92. The clone was cultivated at the Norwegian Institute for Water Research under defined laboratory conditions.[7,8] The verification of toxicity of algal cells was made by means of mouse bioassays.[4,9]

A recently described method for isolation and quantitative determination of alkaloid neurotoxins[10] was used to obtain homoanatoxin extracts from algal cultures and, therefore, the method is only briefly recapitulated: Algal biomass was obtained by filtration of 0.3 l water sample containing cyanophytes with a Whatman GF/C microfiber filter using a Millipore vacuum cruent stand. The biomass retained by the filter was lyophilized. Two ml of 0.05 M acetic acid (MERCK, analytical grade) was added to 5 mg of freeze-dried biomass and ultrasonicated for 5 min. The sample was centrifuged at 15 000 rpm for 30 min. The pH of the aqueous extract was adjusted to pH≥11 with 0.5 M sodium carbonate (Merck, analytical grade) and passed through a C18 cartridge of 1 ml volume (SEP-PAK, Waters) preconditioned with 4 ml of methanol (Rathburn, HPLC grade) followed by 8 ml of distilled water. The sample was applied to the cartridge which was washed with 8 ml of distilled water followed by 8 ml of methanol. The toxin-containing methanol fraction was collected. Twenty µl of the toxin-containing fraction was evaporated to dryness in a reaction vial with a flow of nitrogen. In order to obtain a thermally more stable and less polar compound suitable for gas chromatography, the sample was derivatized as follows: The sample residue was dissolved in 150 µl of acetonitrile (Rathburn, HPLC grade) and heptafluorobutyric acid anhydride (Pierce) (3:1 by volume) was added as acylation reagent. Derivatization was performed at 50±1°C for 20 min. The sample was evaporated to dryness with a flow of nitrogen and redissolved in 200 µl cyclohexane (Rathburn, HPLC grade).

An HP-5980 gas chromatograph connected to an HP-5987A quadrupole mass spectrometer was used for detection. Separation was carried out on a 25 m x 0.2 mm i.d. fused silica column coated with 0.11 µm HP Ultra 2 stationary phase. Helium was used as carrier gas at a flow rate of 30 cm/s. One µl sample was injected splitless at an injector temperature of 250°C, and the following temperature program was employed: 100°C for 2 min, then 10°/min to 250°C, isothermal for 5 min. Full-scan spectra were recorded from m/z 50 to 400 at a scan rate of 600 amu/s. The mass spectrometer was operated in the negative ion chemical ionization (NICI) mode with methane (Messer Griesheim, 99.95% purity) as reagent gas. The ion source pressure was 0.5 bar, and the ion source temperature was 200°C. The electron energy of the primary electrons was in the order of 120 eV.

For enantiomer separation, a glass capillary column of 12 m x 0.32 mm i.d. was used coated with a 3+1 (w/w) mixture of PS086 and permethylated heptakis (2,3,6-tri-O-methyl)-β-cyclodextrin (PMCD) with 0.3 µm film thickness. Further details of its preparation have been described previously.[11] The column temperature program started at 60°C (2 min) and was increased at a rate of 20°C/min up to 120°C followed by a rate of 3°C/min to 200°C (5 min). Helium was used as carrier gas at a flow rate of 35 cm/s.

Racemic synthetic homoanatoxin was used to study the enantiomer separation performance of the chiral GC column. The hydrochloride salt of (+)-anatoxin (Biometric Systems Inc. Minnesota, USA) was used as internal standard for the quantification of homoanatoxin-a. It was added to *Oscillatoria* NIVA-CYA 92 samples prior to sample clean-up. Quantification of the amount of toxin present in a sample was achieved by means of multilevel calibration based on peak area ratios of (+)-anatoxin-a and synthetic (±)-homoanatoxin.

3 RESULTS AND DISCUSSION

The negative ion chemical ionization spectrum of the N-HFB derivative of homoanatoxin-a shows the molecular ion at m/z 375 as well as abundant fragment ions at m/z 355, m/z 335, m/z 315 and m/z 295 (Figure 2). The fragmentation is characteristic for the heptafluoro-butyryl group, which promotes a series of HF eliminations leading to [M-HF]⁻·, [M-2HF]⁻·, [M-3HF]⁻· and [M-4HF]⁻· ions. The base peak at m/z 315 was chosen for selected ion monitoring of homoanatoxin enantiomers.

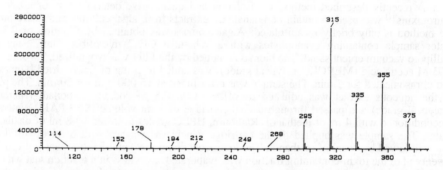

<u>Figure 2</u> Full scan negative ion chemical ionization spectrum of N-HFB derivative of homoanatoxin-a.

For the present, chiral stationary phases for GC are in general less stable than normal polysiloxanes. Therefore, they are more vulnerable to interactions. The presence of active matrix compounds interacting with the stationary phase may affect the enantioselective sites leading to distorted peak shapes and poor separation. In severe cases the column may be damaged or the lifetime of the column is shortened. This requires an effective sample clean-up procedure. Prior to chiral gas chromatography the samples were analyzed on a normal phase GC column (see experimental). Homoanatoxin-a was the dominating signal of the chromatogram. No matrix or background interference could be observed. The injection of a sample containing about 100 pg homoanatoxin-a gave a signal to noise ratio of at least 100:1. Accordingly, the applied clean-up procedure removes the sample matrix efficiently and fulfils the requirements concerning extract purity.

At present it is not possible to predict the exact enantiomer selective behaviour of a chiral stationary phase. Therefore, enantiomer separation by high resolution gas chromatography is still based on an empirical selection of the stationary phase. Weak interactions like van der Waals forces play an important role in the enantiomer separation on modified cyclodextrins. Such forces decrease with increasing temperature, and, consequently, temperature programming should start at low temperatures using slow program rates (see experimental). As Figure 3 shows, the racemic mixture was completely separated within 15 minutes on a glass capillary column as short as 12 m (0.32 mm id.) coated with 20% of a permethylated β-cyclodextrin dissolved in 85%-methyl-15%-phenylpolysiloxane. The column performance was: R=1.56, α=1.025, K=13.44, and N=14425. Accordingly, the stationary phase should also be able to separate enantiomers of compounds related to homoanatoxin.

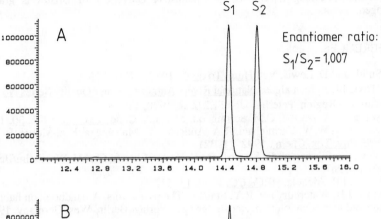

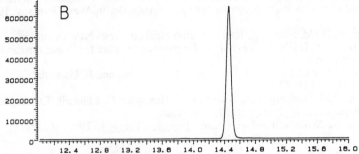

<u>Figure 3</u> Enantiomer selective separation of racemic homoanatoxin obtained by synthesis (A) and homoanatoxin-a from a water extract of *Oscillatoria formosa* NIVA-CYA 92 cell culture (B). Mass m/z 315 [M-3HF]⁻ was used to monitor the compounds in the negative ion chemical ionization mode. The enantiomer ratio of the racemate is marked.

Figure 3 shows the result of the enantioselective separation of about 100 pg of homoanatoxin-a which was isolated from the toxic cell culture of *Oscillatoria formosa* NIVA-CYA 92 as described earlier. The sample had a homoanatoxin-a concentration of 1.9 μg/mg on algal mass dry weight basis. It consists of the first eluting enantiomer exclusively. Since the toxicity of the cyanophyte is obviously only caused by this enantiomer, a remaining question is the biological activity of the other enantiomer present in the racemate. At the moment no further information is available to assign the absolute enantiomer configuration of the biogenically produced compound.

4 CONCLUSIONS

The applied stationary phase separates the homoanatoxin enantiomers completely within 15 minutes. It shows very low stationary phase bleeding. Therefore, it is compatible with selective detectors such as the electron capture detector or NICI mass spectrometry. The presented method allows the enantioselective quantification of sub-pg amounts corresponding sub-ppb levels in water samples. It is assumed that the presented methodology can be applied to quantify other chiral alkaloid neurotoxins produced by algae.

ACKNOWLEDGEMENTS

We thank Olav M. Skulberg at the Norwegian Institute for Water Research for providing sample material from his culture collection of algae. The receipt of racemic homoanatoxin from Dr. Timothy Gallagher, School of Chemistry, University of Bristol is gratefully acknowledged.

REFERENCES

1. R.A. Smith and D. Lewis, Vet. Hum. Toxicol., 1987, 29(2), 153.
2. NRA.'Toxic blue-green algae.' National Rivers Authority,Water Quality Series No. 2. NRA Anglian Region, Peterborough PE2 0ZR, 1990, 128 pp.
3. C. Edwards, K.A. Beattie, C.M. Scrimgeour and G.A. Codd, Toxicon, 1992, 30, 1165.
4. O.M. Skulberg, W.W. Carmichael, R.A. Andersen, S. Matsunaga, R.E. Moore, R. Skulberg, Env. Tox. Chem., 1992, 11, 321.
5. K. Mislow, 'Einführung in die Stereochemie', Verlag Chemie GmbH, Weinheim/Bergstr., Darmstadt, 1967.
6. W. Blum and R. Aicholz, HRC&CC, 1990, 13, 515.
7. R. Rippka, J.B. Waterbury and R.Y. Stainer, 'The prokaryots. A handbook on habitats, isolation and identification of bacteria', Springer-Verlag, Berlin, West Germany, 1981, p. 212.
8. R. Skulberg and O.M. Skulberg, 'Research with algal cultures - NIVA's culture collection of algae', ISBN 82-577-1743-6, Norwegian Institute for Water Research, Oslo, Norway, 1990.
9. K. Berg, W.W. Carmichael, O.M. Skulberg, Chr. Benestad and B. Underdal, Hydrobiol., 1987, 144, 97.
10. J.E. Haugen, O.M. Skulberg, R.A. Andersen, J. Alexander, G. Lilleheil, T.Gallagher and P.A. Brough, Arch. Hydrobiol, 1994, in press.
11. M.D. Müller, M. Schlabach and M. Oehme, Env. Sci. Technol., 1992, 26, 566.

Neurotoxins from Australian *Anabaena*

D. A. Steffensen,[1] A. R. Humpage,[1] J. Rositano,[1] A. H. Bretag,[2]
R. Brown,[3] P. D. Baker,[1] and B. C. Nicholson[1]

[1]AUSTRALIAN CENTRE FOR WATER QUALITY RESEARCH, PRIVATE MAIL BAG,
SALISBURY, SOUTH AUSTRALIA 5108

[2]UNIVERSITY OF SOUTH AUSTRALIA, ADELAIDE, SOUTH AUSTRALIA 5000

[3]DEPARTMENT OF COMMUNITY & HEALTH SERVICES TASMANIA, GPO BOX 125B,
HOBART, TASMANIA 7001

INTRODUCTION

In Australia, toxicity has been associated with 4 genera of cyanobacteria. Hepatotoxicity has been demonstrated from the well known peptides produced by *Nodularia spumigena* and *Microcystis aeruginosa*[1,2] and the alkaloid produced by *Cylindrospermopsis raciborskii* [3]. The agent(s) responsible for the toxicity of *Anabaena* has been less clear. There is some evidence of hepatotoxicity in chickens and mice treated with *Anabaena circinalis*[4,5]. However, the symptoms most consistently associated with *Anabaena* toxicity indicate a neurotoxin[6].

Neurotoxicity associated with freshwater cyanobacteria has been extensively reported throughout the world from a number of taxa[7]. The most commonly identified agent has been anatoxin-a which has been isolated from *Anabaena flos-aquae*[8], *A.circinalis, A.lemmermannii*[9,10], *Aphanizomenon*[5,9] and *Oscillatoria* sp[9,11]. Anatoxin-a is a small alkaloid compound which acts as a post-synaptic cholinergic depolarising agent[12]. The symptoms include muscle fasciculation, loss of coordination, gasping, convulsions and death by respiratory arrest. An analogue of anatoxin-a with similar toxicity has been isolated from *Oscillatoria formosa*[13]. The less common anatoxin-a(s) has also been isolated from *Anabaena flos-aquae*[14,15]. This is a phosphate ester which acts as an acetylcholinesterase inhibitor. Symptoms include marked salivation, lachrymation, urinary incontinence, muscular weakness, fasciculation, convulsions and death by respiratory failure. In a number of cases neurotoxicity has been associated with *Anabaena* without evidence of anatoxin-a or anatoxin-a(s)[9,14,16].

Neurotoxicity has also been associated with blooms of *Aphanizomenon flos-aquae* in New Hampshire[17]. Subsequent investigations indicated that these "aphantoxins" included saxitoxin and neosaxitoxin normally associated with paralytic shellfish poisons (PSPs) from marine dinoflagellates[18-20]. The PSPs are sodium channel blockers which produce gross symptoms similar to those for anatoxin-a.

The results described here are from extensive surveys of the Murray-Darling River Basin in Australia from 1990 to 1993. The study covered 1 million sq km of South Eastern Australia or 14% of Australia.

METHODS

Sample Collection and Processing

Samples were taken by plankton net or grab samples of surface scums depending on the density of the bloom. Subsamples were preserved in Lugol's iodine for cell counts with the remainder stored at -20°C.

Extracts for toxicity testing were prepared from lyophilised material reconstituted in physiological saline and ultrasonicated on ice.

Toxin Determination

Mouse bioassays were conducted by intra-peritoneal injection of 1 ml of extract into pairs of female white Balb/c mice of 16-23 gm. Samples that tested positive at an initial screening dose of 5-10 mg lyophilised cells/ml were retested at progressive dilutions (2.5, 1.25, 0.625 mg ml^{-1} *etc*) to determine the minimum lethal dose. Samples were described as hepatotoxic or neurotoxic depending on the survival time, symptoms and post-mortem examinations.

Anatoxin-a was analysed for by gas chromatography with electron capture detection[21] or with mass spectrometric selective ion monitoring.

The presence of microcystins and nodularin was determined by reverse phase high performance liquid chromatography (HPLC) with UV photodiode - array detection[22].

The presence of PSPs was determined by the HPLC methods described by Oshima *et al.* [23] or Sullivan and Wekell [24]. Fast atom bombardment mass spectrometry (FAB/MS) was also used to support the presence of PSP.

The electrophysiological effect of partly purified extract was tested on the isolated desheathed sciatic nerve of the toad *Bufo marinus*[25].

RESULTS

Of the 231 samples from 130 sites, 98 (42%) were toxic by mouse bioassay. Hepatotoxicity was evident in 42 samples (18%). These samples were dominated by either *M.aeruginosa*, *N.spumigena* or *C.raciborskii*. Neurotoxicity was evident in 56 samples (24%) with all the neurotoxic samples containing *A.circinalis*.

Extensive analysis failed to detect any evidence of anatoxin-a.

The effect of a partly purified extract of a scum of *A. circinalis* on the isolated desheathed sciatic nerve of the toad was a time and dose dependent decrease in the action potential peak height which was identical to that obtained using pure saxitoxin. This indicated that the *A. circinalis* extract most likely acted as a sodium channel blocker.

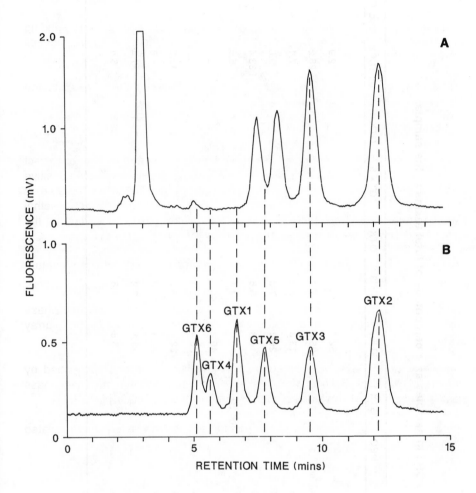

Figure 1. HPLC analysis of gonyautoxins. A, sample from the River Darling (upstream of Wentworth), 28/07/92. The trace clearly shows the presence of GTX 2 and GTX 3 in the sample, and two peaks, one either side of the GTX 5 elution time, which are probably the decarbamoyl derivatives of these toxins (23). B, standard gonyautoxins 1-6. Samples were extracted into 0.1N HCl with sonication, centrifuged and filtered, and then diluted 1/10 in 0.05 N acetic acid. Ten μl was injected onto a PRP-1 column (Hamilton Corp.) and run as per Ref 23.

Table 1. Proportion of PSPs in samples as a percentage of total toxins in the sample.

Sample Code	STX	NEO	GTX 1	GTX 2	GTX 3	GTX 4	GTX 5	GTX 6	dcGTX 2	dcGTX 3
ANA-118 C	5.7	–	–	62.7	15.9	–	–	–	12.9	2.7
BURR	24.3	–	–	35.8	7.2	–	–	–	28.2	4.5
CHAF	73.1	–	–	13.8	3.2	–	–	–	7.1	2.8
HOPE	56.2	–	–	26.2	7.3	–	–	–	9.8	0.8
JANE	31.5	–	–	16.8	5.6	–	–	–	33.4	12.7
LOK2	3.6	–	30.0	15.0	4.1	–	–	40.1	5.6	1.6
MILL	14.1	–	–	40.2	43.7	–	–	–	–	–
MYPA	37.5	–	–	50.4	12.1	–	–	–	–	–
DAR1	1.6	–	–	22.4	5.7	–	–	57.1	10.2	3.1
DAR2	48.5	–	–	24.3	6.0	11.1	–	–	6.8	3.2
WON1	–	–	–	10.1	3.2	–	–	77.1	7.4	2.2
WON2	0.5	–	–	5.8	1.8	–	–	83.3	6.4	2.2

HPLC analysis identified a range of PSPs. As shown in Figure 1 and Table 1 there was evidence of saxitoxin (GTX) gonyautoxins 1, 2, 3, 4 and 6. There was also evidence of the decarbamoyl derivatives dc GTX 2 and dc GTX 3.

The FAB/mass spectra provided supporting evidence for the presence of saxitoxin, neosaxitoxin and dc GTX 2 and dc GTX 3.

DISCUSSION

Neurotoxicity within the study area was exclusively associated with *A. circinalis*. There was no evidence of neurotoxicity associated with other species of *Anabaena* including *A. flos-aquae* and *A. spiroides* or with *Aph. gracile* and *Aph. issatschenkoi*. *Aph. flos-aquae* was not detected during this study.

Although there was no evidence of acute hepatotoxicity from the extracts of *A. circinalis*, post-mortem examinations revealed that some mice had darkened, but normal sized, livers suggesting sub-acute damage. Darkened livers were also present in some mice subjected to extracts of other species which did not cause any gross toxicity symptoms. These included *A.aphanizomenoides, A. flos-aquae, A. solitaria, A. spiroides, Aph. gracile* and *Planktothrix (Oscillatoria) mougeoti*. HPLC analysis of the *A. circinalis* material was negative for microcystin and nodularin. The significance and cause of the darkened livers warrants further investigation.

The neurotoxic effects from *A. circinalis* appear to be accounted for by the PSP detected. The quantities of toxins found were comparable with those purified from *Aph. flos-aquae*[15] but a much wider range of PSPs were identified.

The widespread occurrence of PSPs within Australian *A. circinalis* and unexplained neurotoxic effects from *Anabaena* species elsewhere suggests that more regular analysis for these compounds in neurotoxic blooms is warranted.

REFERENCES

1. M.T.C. Runnegar, A.R.B. Jackson, and I.R. Falconer. *Toxicon* 1988 **26**, 143.

2. D.J. Flett, B.C. Nicholson and M.D. Burch, Proceed. 14th Federal Convention AWWA 17, 1991.

3. I. Ohtani, R.E. Moore, and M.T.C. Runnegar. *J. Am Chem Soc.* 1992 **114**, 7941.

4. E.J. McBarron, R.I. Walker, I. Gardener, and K.H. Walker. *Aust. Vet. J.* 1975 **51**, 586.

5. L.C. Bowling, In Rept. No 92.074 Tech. Serv. D.W. NSW Dept. Water Resources, 1992.

6. M.T.C. Runnegar, A.R.B. Jackson, and I.R. Falconer. *Toxicon* 1988 **26**, 599.

7. W.W. Carmichael, *A Status Report on Planktonic Cyanobacteria (Blue-Green Algae) and Their Toxins.* (US Environmental Protection Agency (EPA/600/R-92/079), Cincinnati, 1992).

8. J.P. Devlin, O.E. Edwards, P.R. Gorham, N.R. Hunter, R.K. Pike, B. Stravic, *Can. J. Chem.* 1977, **55**, 1367.

9. K. Sivonen, K. Himberg, R. Luukkainen, S.I. Niemelä, G.K. Poon, G.A. Codd, *Tox. Assess.* 1989 **4**, 339 (1989).

10. D.K. Stevens and R.I. Krieger, *Toxicon* 1991 **29**, 134-138.

11. C. Edwards, K.A. Beattie, C.M. Scrimgeour, G.A. Codd, *Toxicon* 1992 **30**, 1165.

12. W.W. Carmichael, D.F. Biggs, M.A. Peterson, *Toxicon* 1979 **17**, 229.

13. O.M. Skulberg, W.W. Carmichael, R.A. Anderson, S. Matsunaga, R.E. Moore, R. Skulberg, *Environ. Toxicol. Chem.* 1992 **11**, 321.

14. W.W. Carmichael and P.R. Gorham. *Mitt. Internat. Verein. Limnol.*1978 **21**, 285.

15. N.A. Mahmood and W. W. Carmichael, *Toxicon* 1986 **24**, 175.

16. M. Ekman-Ekebom, M. Kauppi, K. Sivonen, M. Niemi, and L. Lepisto. *Environ. Toxicol and Water Qual.* 1992 **7**, 201.

17. P.J. Sawyer, J.H. Gentile, J.J. Sasner, Jr., *Can. J. Microbiol.* 1968 **14**, 1199.

18. M. Alam, Y. Shimuzu, M. Ikawa, and J.J. Sasner. Jr. *Environ. Sci. and Health.* 1978 **A13(7)**, 493.

19. M. Ikawa, K. Wegener, T.L. Foxall, J.J. Sasner, Jr., *Toxicon* 1982 **20**, 747.

20. J.J. Sasner, Jr., M. Ikawa, T.L. Foxall, in *Seafood Toxins,* E.P. Ragelis, Ed. (American Chemical Society, Washington DC, 1984). p. 391.

21. D.K. Stevens and R.I. Krieger. *J. Analyt. Toxicol.* 1988 **12**, 126.

22. D.J. Flett and B.C. Nicholson, "Toxic Cyanobacteria in Water Supplies: Analytical Techniques". Urban Water Research Association of Australia, Research Report No 26 1991.

23. Y. Oshima, K. Sugino, T. Yasumoto, in *Mycotoxins and Phycotoxins '88*, S. Natori, K. Hashimoto, Y. Ueno, Eds. (Elsevier Science Publishers B.V., Amsterdam, 1989). p 319.

24. J.J. Sullivan and M.M. Wekell, in *Seafood Toxins,* E.P. Ragelis, Ed. (American Chemical Society, Washington DC, 1984). p. 197.

25. C.Y. Kao, Pharmacol. Rev 1966 **18**, 997.

The Analysis of Microcystin-LR in Water: Application in Water Treatment Studies

H. A. James, C. P. James, and J. Hart

WRC PLC, HENLEY ROAD, MEDMENHAM, MARLOW, BUCKINGHAMSHIRE
SL7 2HD, UK

1 INTRODUCTION

Blooms of blue-green algae (cyanobacteria) in raw water storage reservoirs have affected the operation of water treatment plants and caused taste and odour problems in treated water for many years. The water supply industry in the UK has been concerned that toxins released by blue-green algae could contaminate potable supplies. The potential problems were highlighted in the late 1980s by water resources difficulties associated with low rainfall and reports of toxin-induced illness in animals and humans using recreational waters.

Accordingly WRc was commissioned to undertake a comprehensive programme of research which would provide guidance to water treatment plant operators on safeguarding the quality of drinking water supplies. Some aspects of this work (funded by the Foundation for Water Research, the Department of the Environment and the National Rivers Authority) are discussed in relation to the occurrence of microcystin-LR in raw waters and the effectiveness of removal by water treatment processes.

Information available when this work commenced in 1990 indicated that the toxin of most concern was the hepatotoxin microcystin-LR, as it was known to be the most commonly encountered microcystin[1] and one of the most toxic. Another important consideration was the commercial availability of high purity standards - at the time, this was the only toxin obtainable in sufficient quantities to allow a comprehensive programme of research to be undertaken.

2 DEVELOPMENT OF AN ANALYTICAL METHOD FOR THE DETERMINATION OF MICROCYSTIN-LR IN WATER

A primary requirement was the development of a sufficiently sensitive method of analysis to determine the levels of the compound of interest in water. Although there were a few reports of the measurement of microcystin-LR in water prior to commencement of work at WRc, little or no information was presented to allow an assessment of their reliability or their performance characteristics (limits of detection, precision and reproducibility). The method development and associated optimisation work undertaken by WRc has been described in detail elsewhere[2].

The fully developed method involves addition of an internal standard (nodularin) to the water sample (250 ml), filtration, a clean-up step using a solid phase cartridge (Bond Elut^R Aminopropyl), solid phase extraction (Bond Elut^R CBA), followed by HPLC analysis with UV detection. A post column fluorescence derivatisation was investigated. Although it was expected that this would increase the specificity of detection, as the reaction involved only derivatised arginine-containing peptides, this was not the case as it was found that a fluorescence response (at the excitation/emission wavelengths used to detect the derivative) from material which occurs in natural waters gave an unacceptably high background.

A limited evaluation (four levels in the range 1 - 50 µg l^{-1} in reservoir water, four replicates at each level; five levels in the range 0.2 - 50 µg l^{-1} for reservoir-derived drinking water, four replicates at each level) of the method with UV detection demonstrated that the performance of the method was acceptable, with coefficients of variation for drinking water being less than 10% at levels of 1 µg l^{-1} and above.

Performance Testing of the Method

Following the successful development of a method for the analysis of microcystin-LR in water, a decision was taken by the Foundation for Water Research and the Department of the Environment to fund rigorous testing of the performance of the method. A general invitation was issued to interested laboratories to tender for a specified work programme, and five laboratories were chosen to participate. Eleven duplicate batches of samples were analysed at each of four levels *viz.* drinking water spiked at 0.2 and 4.0 µg l^{-1}, reservoir water spiked at 0.3 and 8 µg l^{-1}. The lower levels were chosen as probable lowest detectable levels, from which a statistically derived limit of detection could be calculated. Calibration lines were established for the range 0 - 10 µg l^{-1} by each laboratory. In all, each laboratory analysed at least 112 samples. One laboratory reported difficulties in detecting microcystin-LR at the lowest levels. This appeared to be due to the fact that a computing integrator rather than a PC-based data handling system was used. The latter generally provides better data manipulation facilities, which assist with the detection and quantification of low responses.

The results obtained are shown in Table 1. At low levels of microcystin-LR (0.2 µg l^{-1} in drinking water; 0.3 µg l^{-1} in reservoir water) the relative standard deviation (RSD) was in the range 17 - 52%, while at higher levels (4 µg l^{-1} in drinking water; 8 µg l^{-1} in reservoir water) the RSD was in the range 5 - 30%, with three of the laboratories achieving an RSD of 10% or less. This shows that the initial assessment of the performance of the method during its development was realistic.

The method is currently being considered by the Standing Committee of Analysts for adoption as one of the series of Methods for the Examination of Waters and Associated Materials.

3 PERSISTENCE OF MICROCYSTIN-LR

The persistence of microcystin-LR released from algal cells (e.g. during cell senescence and lysis) is an important factor in relation to drinking water supply and recreational use of waters affected by blooms or high algal cell densities. For example, if the toxin was only slowly biodegraded following release from algal cells, a significant algal bloom could result in the suspension of recreational activities in an affected water body for prolonged periods, and there might well be serious financial consequences. Similar problems would arise if additional treatment processes had to be applied to safeguard drinking water supplies.

The method developed to determine microcystin-LR in water was applied to monitor the disappearance of microcystin-LR from reservoir water under varying conditions. It was shown that in sterilised reservoir water, or in groundwater, microcystin-LR is persistent. However in unsterilised reservoir water, particularly when bed sediment was also present, microcystin-LR was biodegraded (Figure 1) with a half-life under the experimental conditions employed of 3 - 4 days. Similar results have recently been reported by other workers[3]. Depending on the initial levels of microcystin-LR present, and the levels which would give rise to concern, it is therefore possible to estimate the length of time water bodies are likely to be affected.

Table 1. Results from multi-laboratory performance testing of method for microcystin-LR in drinking water and reservoir water*

Laboratory	Drinking water							Reservoir water						
	Level+ (μg l^{-1})	Mean+ (μg l^{-1})	S_w	S_b	S_t	RSD(%)	Deg. of F	Level+ (μg l^{-1})	Mean+ (μg l^{-1})	S_w	S_b	S_t	RSD(%)	Deg. of F
1	0.2	0.2277	0.1062	0.0000	0.1062	47	20	0.3	0.2941	0.1541	0.0000	0.1541	52	21
2	0.2	0.2531	0.0336	0.0258	0.0424	17	16	0.3	0.2480	0.0239	0.0365	0.0436	18	12
3	0.2	0.1256	0.0330	0.0643	0.0723	58	10	0.3	0.3472	0.0576	0.1121	0.1260	36	10
4	0.2	0.4230	0.0689	0.0123	0.0700	17	21	0.3	0.4236	0.0626	0.0000	0.0626	15	17
5	0.5	0.7971	0.0689	0.0586	0.0904	11	17	1.0	0.7535	0.1027	0.0747	0.1269	17	18
1	4.0	3.7464	0.3312	0.0616	0.3368	9	21	8.0	7.6591	0.6287	0.0000	0.6287	8	21
2	4.0	4.0476	0.4029	0.0000	0.4029	10	16	8.0	8.3089	0.8125	0.2489	0.8498	10	18
3	4.0	3.6565	0.5309	0.9793	1.1140	28	11	8.0	7.8225	0.8074	1.5710	1.7664	22	13
4	4.0	3.8036	0.2913	0.2510	0.3846	10	17	8.0	8.2795	0.2583	0.3346	0.4227	5	14
5	4.0	4.2605	0.6756	0.1804	0.6993	16	21	8.0	7.6298	1.9313	1.2372	2.2936	30	19

+ Level indicates the level at which the samples were spiked; Mean indicates the mean level found.

* The laboratories taking part in the performance testing analysed 11 batches of duplicate samples at the levels indicated. The participating laboratories were: LabServices, Analytical and Environmental Services, The Water Quality Centre (Thames Water), Bath University (School of Pharmacy and Pharmacology) and WRc, though not necessarily in the order indicated.

S_w Within batch standard deviation
S_b Between batch standard deviation
S_t Total standard deviation
RSD Relative standard deviation (coefficient of variation)
Deg. of F Degrees of freedom

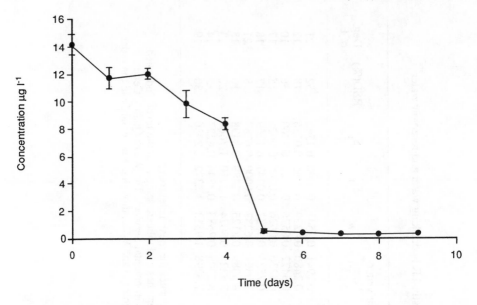

Figure 1. Degradation of microcystin-LR in reservoir water.

4 EFFECTIVENESS OF TREATMENT PROCESSES IN REMOVING MICROCYSTIN-LR

Laboratory and pilot scale studies had shown that conventional treatment processes such as chemical coagulation, sand filtration and chlorination were ineffective in removing microcystin-LR from raw waters[4]. Therefore advanced water treatment processes were investigated in a series of laboratory tests. These were activated carbon adsorption (both powdered (PAC) and granular (GAC)), oxidation (using ozone or other chemical oxidants) and membrane filtration.

The water used for the experiments was either raw water taken from a lowland reservoir source (pH 7.7, turbidity 0.8 NTU, TOC 5.7 mg l^{-1}, 6 °Hazen true colour) or from the same source after treatment by clarification and rapid sand filtration (pH 7.9, turbidity 0.4 NTU, TOC 4.2 mg l^{-1}, 5 °Hazen true colour). Microcystin-LR, at concentrations of up to 10 µg l^{-1}, was spiked into the water samples used.

Powdered Activated Carbon

To assess the effect of treatment with PAC, standard coagulation jar test procedures were used to simulate PAC dosing into a floc-blanket clarifier, in conjunction with ferric sulphate as coagulant. The procedure used consisted of coagulant and PAC addition and rapid mixing, flocculation and settlement. Three types of PAC were tested with a range of doses. Each PAC was derived from a different source material *viz.* coal (Chemviron GW), wood (Pica Picazine) and coconut (Sutcliffe Speakman 207CP).

The results obtained (Figure 2) demonstrate that the coal and coconut PACs gave similar, but poorer, removal than the wood-based PAC. A dose of 20 mg l^{-1} of the latter resulted in 85% removal of microcystin-LR, compared to only about 40% removal for the other two types.

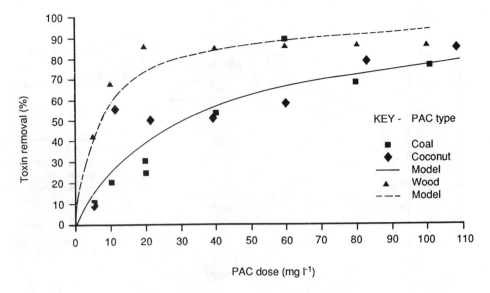

Figure 2. Removal of microcystin-LR with different PAC types.

Granular Activated Carbon

To assess the effectiveness of GAC adsorption, the rapid column test (RCT) procedure[5] was employed. RCTs mimic full scale adsorbers. GAC ground to a small particle size is used in a short column, with a contact time of only a few seconds. The reduction in particle size increases the rate of adsorption, thus allowing RCTs to be conducted on a much reduced timescale compared to full scale or pilot scale studies. For this work, the RCTs were operated to simulate a GAC bed operated with a contact time of 6 minutes for a period equivalent to 12 months.

Four RCT studies were conducted to compare four different types of GAC, which were chosen to represent four commonly used source materials - coal (Chemviron Filtrasorb 400), peat (Norit PKO.6-2), wood (Pica Picabiol H120) and coconut (Sutcliffe Speakman 207C).

All four GACs removed microcystin-LR. However, as can be seen in Figure 3, breakthrough of the toxin was rapid. The coconut GAC gave significantly poorer removal than the other types, with 80% breakthrough occurring after treatment of 5,000 bed volumes of water. With the other three types, 80% breakthrough occurred after 30,000 bed volumes (equivalent to 18 weeks of operation).

These results show that with a short contact time of six minutes and a persistent toxin problem, it would be difficult to obtain long GAC bed lives. However, these studies cannot take into account any biological activity on the GAC. Microcystin-LR has been shown to be biodegradable and slow sand filtration has also been shown to remove toxins produced by *Microcystis* sp.[4]. With longer contact times and only seasonal occurrences of microcystin-LR, it is considered unlikely that breakthrough of this toxin will occur as it will probably be biodegraded on the GAC.

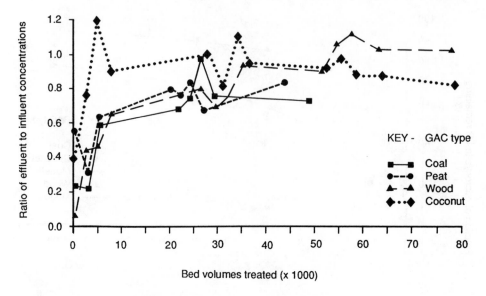

Figure 3. Removal of microcystin-LR with different GAC types.

Oxidation

To assess the effect of various oxidants, experiments with ozone, chlorine, chlorine dioxide, hydrogen peroxide and potassium permanganate have been carried out with both raw and clarified waters.

For the ozonation experiments, a laboratory scale bubble-diffuser contacting system was used, with a range of ozone doses up to 9.2 mg l^{-1} and contact times of 5 - 10 minutes. Typical results are shown in Table 2, from which it can be seen that microcystin-LR was almost totally removed in raw and clarified waters at ozone doses of about 2.5 mg l^{-1}.

Table 2. Removal of microcystin-LR from water using ozone

Water type	Ozone dose (mg l^{-1})	% Removal
Raw	0.7	12
	1.8	24
	2.5	>96
Clarified	1.2	73
	2.4	>99
	3.6	>99

Of the other oxidants tested, both chlorine and hydrogen peroxide were shown to be ineffective in removing microcystin-LR from either raw or clarified waters. Chlorine dioxide was ineffective when applied to raw water, but resulted in significant removal when applied to clarified water at high doses (removal >80% with dose levels >6 mg l^{-1}). Potassium permanganate was the most effective of these other oxidants, accomplishing significant toxin removal when applied to both raw and clarified waters (Figure 4). As with chlorine dioxide, it was most effective when applied to clarified water.

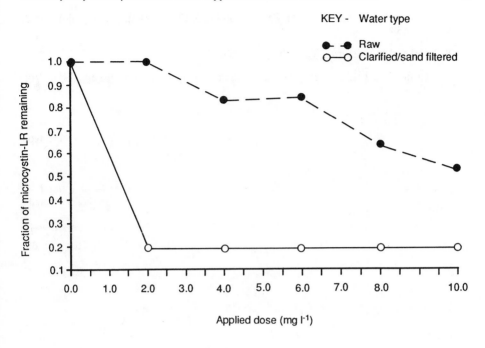

Figure 4. Effect of water type on microcystin-LR removal by potassium permanganate.

Membrane filtration

The effectiveness of a membrane process was assessed using a Film Tech N70 polymeric nanofiltration membrane. This has a nominal molecular weight cut-off of 200. The experiments were conducted with a permeate-to-retentate ratio of 0.25, and the results showed that there was complete rejection of microcystin-LR.

5 CONCLUSIONS

A reliable method has been developed for the analysis of low levels of microcystin-LR in water. This method has been used to demonstrate that this toxin is biodegraded in reservoir water, and to assess the effectiveness of various advanced treatment processes in removing microcystin-LR during water treatment. Both powdered activated carbon and granular activated carbon are effective. Of the oxidative processes investigated, ozonation and treatment with potassium permanganate were both particularly effective. The latter oxidant may offer the most flexible approach to microcystin-LR removal as it can be readily implemented at any site with virtually no capital costs, and running costs are relatively low.

References

1. NRA. 'Toxic blue-green algae. National Rivers Authority, Water Quality Series No. 2.' NRA Anglian Region, Peterborough PE2 0ZR, 1990, 128 pp.

2. H.A. James and C.P. James, Development of an Analytical Method for Blue-green Algal Toxins. Foundation for Water Research Report No FR 0224. FWR, Allen House, The Listons, Liston Road, Marlow, Bucks. SL7 1FD.

3. S.L. Kenefick, S.E. Hrudey, H.G. Peterson and E.E. Prepas, <u>Wat. Sci. Technol.</u>, 1993, Vol. 27, 433.

4. K. Lahti and L. Hiisvirta, <u>Water Supply</u>, 1989, Vol. 7, 149.

5. J.C. Crittenden, J.K. Berrigan and D.W. Hand, <u>J. Wat. Poll. Control Fed.</u>, 1986, Vol. 58, 312.

The Analysis of Microcystins in Raw and Treated Water

Linda A. Lawton,[1,2] Christine Edwards,[1] and Geoffrey A. Codd[1]

[1]DEPARTMENT OF BIOLOGICAL SCIENCES, UNIVERSITY OF DUNDEE, DUNDEE DD1 4HN, UK

[2]CURRENT ADDRESS: DEPARTMENT OF APPLIED SCIENCES, ROBERT GORDON UNIVERSITY, ABERDEEN AB1 1HG, UK

1. INTRODUCTION

In recent surveys, peptide hepatotoxins, known as microcystins, were shown to be the most commonly detected cyanobacterial toxins (Codd and Beattie, unpublished results). Microcystins are cyclic heptapeptides where the general structure is cyclo(-D-ala-L-X-erythro-β-D-methylaspartic acid-L-Y-Adda-D-isoglutamic acid-N-methyldehydroalanine), where X and Y are variable amino acids such as tyrosine and leucine in microcystin-LY. As well as variations in amino acids there are also minor structural modifications giving rise to analogues of variants[1-3], giving in excess of fifty microcystins. Pentapeptides of similar structure have also been identified, nodularin isolated from *Nodularia spumigena* and recently motuporin isolated from a marine sponge[4,5].

All of these cyclic peptides show similar biological activity in that they cause extensive liver damage, and at the molecular level, they are potent and specific inhibitors of protein phosphatases 1 and 2A, which are essential in cellular function, implicating them as potential tumour promoters[6,7].

The potential number of peptide variants coupled with the lack of a wide range of standard reference material, have posed a problem in the development of suitable analytical methods with the result that many microcystins have gone undetected. This paper describes a method which facilitates the identification and quantification of microcystins within cyanobacterial cells and free in the water. The procedure involves sample filtration to separate cyanobacterial cells from water, allowing intracellular and extracellular peptide levels to be assessed. Data on reproducibility and efficiency of this method have been assessed and are presented elsewhere[8].

2. MATERIALS AND METHODS

During method development, it was neccessary to examine procedures for the rapid extraction of microcystins from cyanobacterial cells, both from filtered water samples and scum material. The efficiency of several commonly used

solvents was evaluated using *Microcystis aeruginosa* PCC 7820
which produces microcystins, -LR, -LY, -LW and -LF[9].
Only a summary of the proposed method is presented as the
protocol is fully described elsewhere[8].

3. RESULTS AND DISCUSSION

Methanol was shown to be the most suitable solvent for
extracting the range of microcystins produced by *M. aeruginosa*
PCC 7820 (Figure 1). Although reasonable recoveries were
obtained with butanol/methanol/water, this solvent system was
found to be less practical.

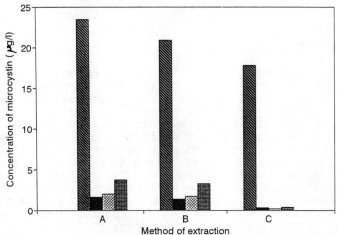

Figure 1 - Extraction of 4 microcystins (▨ MC-LR, ■ MC-LY
▨ MC-LW and ▨ MC-LF) from freeze-dried cells of *M.
aeruginosa* PCC 7820 by A: methanol, B: butanol/methanol/water
(1:4:15) and C: acetic acid (5% v/v)

As well as extracting peptides from cells retained on
filters, methanol can be used to extract cells from blooms (50
mg dry cells per ml methanol). A number of samples from sites
around the UK were extracted in this manner and analysed by
HPLC with diode array detection. Microcystin composition was
found to vary between species and genus, as illustrated in
Figure 2. By most methods, many microcystins would remain
undetected, however, by employing diode array detection these
can be readily identified on the basis of their characteristic
UV spectra.

Liquid chromatography-mass spectrometry (LC-MS) has been
successfully used to identify microcystins in water samples[10].
However, without spectral confirmation its uses are limited
since it would have limitations for monitoring unknowns since
there are often compounds, possibly related to microcystins,
of similar molecular weight which are not microcystins. Only
fragmentation would reveal if microcystins were present. This

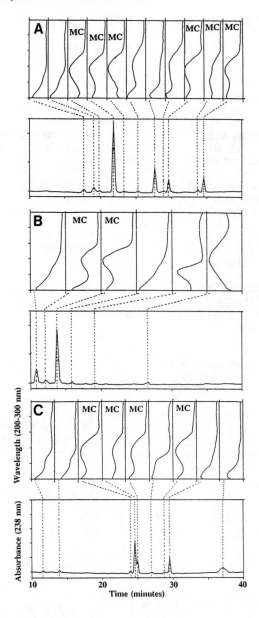

Figure 2 - Detection of microcystins in methanolic extracts of blooms of A: *Microcystis*, B: *Anabaena*, C: *Oscillatoria* based on characteristic UV spectra with absorbance maxima at 239 nm and where tryptophan is included at 222 nm (*).

EXTRACTION OF MICROCYSTINS/NODULARINS FROM WATER/CELLS

2 x 1l water samples
filtered through GF/C

Add sodium sulphite
(100 μl of 1 g/100 ml)

Add internal standard
to 1l (e.g.100 μl of
50 μg/ml microcystin-LR)

Each 1l sample is split
into 2 x 500 ml and
acidified with TFA (5 ml
of 10%)

Each sample is filtered
through GF/C filter

Each 500 ml sample is
applied to a preconditioned
C18 cartridge using a vacuum
manifold system

Cartridges are washed with
10 ml of 10, 20 and 30%
(v/v) aqueous methanol

Air is drawn through for
30 min

Microcystins/nodularins eluted
3 ml acidified methanol
(0.1% v/v TFA)

Samples dried under N_2 at 45°C

Resuspend in 2 x 100 μl methanol

Redried in sample concentrator
and resuspended in 75 μl 70%
aqueous methanol.

Samples analysed by reverse
phase HPLC with diode
array detection (25 μl injection)

Cells retained on filter
are freeze/thawed

Filters extracted 3 times
in 100% methanol (20 ml)

Samples dried *in vacuo* at
40°C

Resuspend in 2 x 250 μl
methanol

Inject 25 μl onto HPLC

is more time consuming, and more useful as a confirmatory technique rather than a routine monitoring technique.

In conclusion, the combination of extraction/concentration procedures described followed by analysis using reverse-phase HPLC with diode array detection provides a versatile method for identification and quantification of a wide range of microcystins/nodularins in the water environment. This method is a valuable tool for research and in environmental monitoring programmes, providing more in depth information on the occurrence and significance of microcystins/nodularins.

REFERENCES

1.   Santikarn, S., Williams, D.H., Smith, R.J., Hammond, S.J., Botes, D.P., Tuinman, A.A., Wessels, P.L., Viljoen, C.C. and Kruger, H. J. Chem. Soc. Chem. Commun., 1983, 275, 652.
2.   Botes, D.P., Tuinman, A.A., Wessels, P.L., Viljoen, C.C., Kruger, H., Williams, D.H., Santikarn, S., Smith, R.J. and Hammond, S.J. J. Chem. Soc. Perkin Trans., 1984, 1, 2311.
3.   Carmichael, W.W., Beasley, V.R., Bunner, D.L., Eloff, J.L., Falconer, I.R., Gorham, P.R., Harada, K., Yu, M., Krishnamurthy, T., Moore, R.E., Rinehart, K.L., Runnegar, M., Skulberg, O.M. and Watanabe, M. Toxicon, 1988, 26, 971.
4.   Sivonen, K., Kononen, K., Carmichael, W.W., Dahlem, A.M., Rinehart, K.L., Kiviranta, J. and Niemela, S.I. Appl. Environ. Microbiol, 1989, 55, 1990.
5.   de Silva, D.E., Williams, D.E., Anderson, R.J., Klix, H., Holmes, C.F.B. and Allen, T.M. Tetrahed. Letts., 1992, 33, 1561.
6.   MacKintosh, C., Beattie, K.A., Klumpp, S., Cohen, P. and Codd, G.A. FEBS Letts., 1990, 264, 187.
7.   Nishiwaki-Matsushima, R., Ohta, T., Nishiwaki, S., Suganuma, M., Kohyama, K., Ishikawa, T., Carmichael, W.W. and Fujiki, H. J. Cancer Res. Clin. Oncol., 1992, 118, 420.
8.   Lawton, L.A., Edwards, C. and Codd, G.A. The Analyst, 1994, 119, 1525.
9.   Lawton, L.A., Edwards, C., Beattie, K.A., Pleasance, S., Dear, G.J. and Codd, G.A. (Submitted to Natural Toxins; 1994).
10.   Edwards, C., Lawton, L.A., Beattie, K.A., Codd, G.A., Pleasance, S. and Dear, G.J. Rapid Comm. Mass Spectrom., 1993, 7, 714.

Application of HPLC and Mass Spectrometry (MALDI) to the Detection and Identification of Toxins from *Microcystis, Oscillatoria* and *Aphanizomenon* from Several Freshwater Reservoirs

C. S. Dow, U.K. Swoboda, P. Firth, and N. Smith

DEPARTMENT OF BIOLOGICAL SCIENCES, UNIVERSITY OF WARWICK, COVENTRY
CV4 7AL, UK

There is an increasing frequency of reservoirs and lakes, which are used for recreation and as sources of drinking water, supporting the growth of toxic cyanobacteria. These organisms may be present in fairly large numbers throughout the year or they may predominate only during a cyanobacterial bloom. The isolation, purification and characterisation of toxins produced by naturally occurring cyanobacteria is particularly important both in terms of the presence of the toxins in the cells which may be consumed accidentally and the possible release of these toxins into the water.
This paper reports on the application of high performance liquid chromatography and matrix assisted laser desorption, time of flight mass spectrometry (MALDI) to the qualitative and quantitative characterisation of both isolated toxins and cyanobacterial biomass.

The HPLC procedure described by Dow *et al.*(1992)[1] was used to facilitate the identification and isolation of hepatocyanotoxins from cellular biomass and water concentrates from several freshwater reservoirs in central England. The toxicity of the environmental samples and the peak fractions collected following HPLC were determined by intraperitoneal injection of mice. The limit of detection of the HPLC procedure was established as 10-20ng using UV detection (equivalent to $70\mu g\ l^{-1}$ in raw water) - Pharmacia system - and 20-40ng using a photodiode array system (equivalent to $150\mu g\ l^{-1}$ in raw water) - Varian LC Star System. However, although the latter detection system had reduced sensitivity it presents the ability to (i) perform spectral analysis of the eluted material which enables comparison of unknowns with standard toxins giving an enhanced probability of identification of peptide toxins and (ii) assessment of the purity of eluate peaks. These analyses were performed using the Varian software 'Polyview'. Figure 1 shows the typical high performance liquid chromatogram of a mixture of the toxin standards, microcystin-LR, microcystin-RR and nodularin. These gave excellent spectral matches to individually analysed toxins.

Intraperitoneal injection of cell lysate prepared from cell biomass of *Microcystis aeruginosa* PCC7806 caused the death of mice, the toxic symptoms being characteristic of hepatotoxins. Autopsy of these mice showed engorged liver with internal bleeding. Biomass harvested from the exponential growth phase gave only one toxic peak from HPLC analysis which was indistinguishable from microcystin-LR. However, two spectrally pure toxic peaks, t_R 24.6 and t_R25, were resolved by HPLC of an early stationary phase population (Figure 2). The primary peak gave an excellent match to microcystin-LR and had a relative molecular mass, as determined by MALDI mass spectrometry, of 998 while the second eluate had only a relatively good match to it, indicating that it was probably a microcystin-like derivative. This is in keeping with the report of Dierstein *et al* (1990)[2] who, using an alternative extraction and isolation procedure in conjunction with thin layer chromatography, mass spectrometry and nuclear magnetic resonance techniques, identified two toxic fractions from *M. aeruginosa* PCC7806 as being microcystin-LR and a microcystin-LR derivative.

The isolation technique and HPLC procedure used in this study permits the rapid identification of cyanobacterial hepatotoxins which are microcystin-like, however, more detailed information requires the use of mass spectrometry and ultimately, amino acid analysis.

Figure 1   HPLC eluate profile of hepatotoxin standards - nodularin t_R 23.3, microcystin-LR t_R 26.5 and microcystin-RR t_R 31.04. Resolution of a mixture of 300ng of each toxin dissolved in 30% methanol in a gradient of 30-60% acetonitrile containing 0.05% trifluoroacetic acid at a flow rate of 1ml min^{-1}.

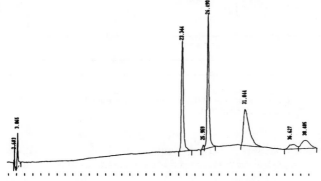

Hepatotoxins were also extracted from environmental biomass and reservoir water samples collected from several sites[3]. HPLC analysis of small scale extractions (HPLC injection equivalent to an extract from 3.3mg dry weight of cells) of a *Microcystis* sp. biomass revealed the presence of a major pure toxic peak with a perfect match to microcystin LR which had a relative molecular mass of 998, as determined by MALDI mass spectrometry. However, large scale extraction (10g dry weight) of the same biomass resulted in the resolution of several toxic peaks all of which had different relative molecular masses (Figure 3) and toxicity levels (Table 1). The relative molecular masses of the toxic fractions (primary MALDI peaks) were 1039, 1050, 1046, 983, 996 and 998 respectively with the peak of molecular mass 998 being the most predominant. The time required to cause death of mice following intraperitoneal injection of equivalent amounts of the toxic eluates varied from 1.5 hours to 9 hours.

Figure 2   Spectral purity analysis (220 - 367nm) of toxic eluates from HPLC analysis of cell biomass of *Microcystis aeruginosa* 7806 in the early stationary phase - two toxic peaks with good similarity to the microcystins were resolved. The second and third peaks are those from the toxic eluates showing similarity to the microcystins. Purity determinations are made by assessing the degree of difference in the UV spectra across the width of the peak. If the peak is free of coeluting compounds the purity parameter will not change over the course of peak elution and will appear with a flat top. If impurities are present the top will slope.

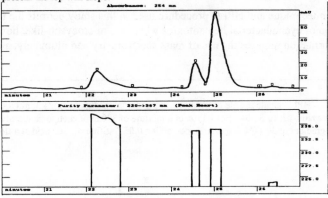

Figure 3   MALDI mass spectrum of the first toxic eluate from *Microcystis* sp. from Cropston.

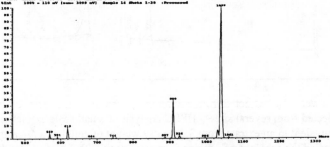

Table 1   Analysis of HPLC toxic eluates from a *Microcystis* bloom in Cropston reservoir (large scale preparation i.e. 10g dry weight of cell biomass)

Eluate	Relative Molecular Mass	Time of Death (intraperitoneal injection)
15	1039	9 - 10 hours
16	1050	2.5 hours
17	1046	1.75 hours
18	983	2 hours
19	996	2 hours
20	998	1.5hours

A toxic strain of *Oscillatoria* sp. was found to be permanently resident in Lower Shustoke reservoir, however, the toxicity of the cell biomass varied throughout the year. This variability has been correlated with the variation in the proportion of the three toxic peptides produced by this strain. These peptides have small structural alterations, as indicated by their relative molecular masses. Details of this study are given in the accompanying paper by Chaivimol *et al.* (1993).

The variability of microcystin-like hepatotoxins from different cyanobacterial strains is not surprising since over 50 such hepatotoxins have been reported to date with several of these having been isolated from the same strain[4-5].

In the course of an environmental monitoring programme several toxic *Aphanizomenon* species were encountered which did not contain hepatotoxins and gave toxic symptoms in mouse bioassays distinct from those of the hepatotoxins. These isolates were therefore assumed to be producing neurotoxins. Consequently a procedure is under evaluation which detects the presence of both hepatotoxins (microcystins) and the neurotoxins structurally related to anatoxin-a in the same sample (Figure 4). This procedure is currently being used to examine toxic environmental samples. However, a toxic eluate has yet to be purified from the *Aphanizomenon* isolates which corresponds to either of these toxin classes.

Figure 4   HPLC eluate profile from a reversed phase C_{18} analytical column of a mixture containing 1µg each of anatoxin a (t_R 14.9), nodularin (t_R 39.8), microcystin-LR and microcystin-RR (coelution t_R 42.8).

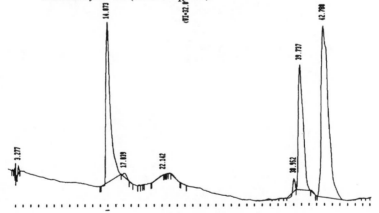

Throughout the environmental monitoring programme cell free water concentrates from bloom situations were assayed for toxicity in mouse bioassays (injection of the equivalent of 10 litres of water) and for the presence of hepatotoxins by HPLC analysis. Despite the sensitivity of the analysis and the proven ability to recover the microcystins from spiked samples (concentrations as low as 170ng/litre of microcystin-LR and nodularin can be readily detected) no 'free' toxins were detected in any reservoir water concentrates, irrespective of the cyanobacterial biomass loading.

REFERENCES

1.   C.S. Dow, U.K. Swoboda and V. Howells, "Recent Advances in Toxinology Research", ed. P. Gopalakrishnakone and C.K. Tan, National University of Singapore, 1992. Vol.3, p323.
2.   R. Dierstein, I. Kaiser, J. Weckesser, U. Matern, W. Knig, and R. Krebber, System. Appl. Microbiol, 1990, 13, p86.
3.   U.K. Swoboda, C.S. Dow and A. Wilson, "Recent Advances in Toxinology Research", ed. P. Gopalakrishnakone and C.K. Tan, National University of Singapore, 1992, Vol.3, p307.
4.   M. Namikoshi, K. Sivonen, W.R. Evans, W.W. Carmichael, F. Sun, L. Rouhiainen, R. Luukkainen, and K.L. Rinehart, Toxicon, 1992, 30, p1457.
5.   W.W. Carmichael, Journal of Applied Bacteriology, 1992, 72, p445.

Routes of Intoxication

R. B. Fitzgeorge, S. A. Clark, and C. W. Keevil

CENTRE FOR APPLIED MICROBIOLOGY AND RESEARCH, PORTON DOWN,
SALISBURY, WILTSHIRE SP4 0JG, UK

1   INTRODUCTION

That algal toxins are involved in episodes of poisoning in
relation to a wide variety of animal species, causing animal
deaths and human illness, is beyond dispute[1].   It is also
apparent that diagnosis is difficult, often based on circumstantial evidence such as an association with expanses of
water containing cyanobacterial blooms together with exclusion of other possible causes.   In addition possibly con-
fusing symptoms of wide diversity are often observed[1].   The
considerable weight of information from the more severe animal
poisoning episodes together with experimental studies give
rise for concern that cyanobacteria may represent a hazard to
human health and as a result requires evaluation.   Therefore,
this present study was devised to examine the effects of algal
toxins on a mammalian animal model in relation to routes and
doses pertinent to the human species.

2   MATERIALS AND METHODS

Two algal toxins were selected for this study, microcystin-LR
to represent a typical algal hepatoxin and anatoxin-a as an
algal neurotoxin, both being available commercially
(Calbiochem Corporation, La Jolla, USA) in a suitably purified form.   The routes of administration of these toxins,
used because they were considered most applicable to the human
species, were:

1.   Oral, which was equated in the animal by gastric
intubation (g.i.).

2.   Inhalation, as represented (a) by administration of a
fine particle aerosol (a.i.) of particle size 3-5μm diameter,
generated by a Collison spray[2] containing 50μg microcystin-
LR/ml, and (b) by intranasal instillation (i.n.), to equate
with inhalation of large particles (>10μm in diameter).

3.   Intraperitoneal   injection,   which was included as a
standard   comparison   route   and   for   bio-assay purposes.

Table 1   Mouse response to administration of microcystin-LR
           by different routes: LD_{50}

Route	LD_{50}
i.p.	250 μg/kg
g.i.	3000 μg/kg
i.n.	250 μg/kg
aerosol inhalation	No deaths at 0.0005 μg/kg

The mammalian animal model chosen was newly weaned CBA/Balbc
mice weighing 20g ($\pm$ 1g).

3   RESULTS

Response to the Administration of microcystin-LR by Different Routes.

LD_{50}.   Where deaths occurred they did so within 2 hrs of
microcystin-LR administration. LD_{50} values were estimated by
the method of Reed and Muench[3].   Table 1 shows the dose of
microcystin-LR required to produce an LD_{50} when administered by
different routes. The results show that the two most sensitive
routes of administration are the intraperitoneal and the
intranasal, both giving an LD_{50} of 250μg/kg.   The most obvious
natural route of exposure is the oral one, equating in this
study to gastric intubation.   However, this route was 12 fold
less sensitive than the other two used.   No illness or histopathological change was observed among those mice receiving
microcystin-LR by aerosol inhalation (exposure to one dose
level only).   However, the dose delivered by this route was
very low (approximately 0.0005μg).   This results essentially
from a combination of the microcystin-LR being aerosolised and
the small lung capacity of the mouse.   To achieve delivery of
a higher dose into the mouse lung would require a much higher
concentration of microcystin-LR in the spray apparatus which
proved to be impracticable in this study.

Weight increase.   Table 2 shows the weight increase of
organs of mice which died after administration of 1 LD_{50} of
microcystin-LR.   The responses were similar irrespective of
route of administration.   No increases in weight of the lungs
and spleen was observed.   In contrast, liver and kidney
weights increased in treated animals by approximately 45% and
8% respectively.

Microcystin-LR: effect of sub-lethal doses administered
to mice by the intranasal route.   Table 3 indicates a dose
response and relationship between amount of microcystin-LR
administered i.n. and the increase in subsequent mouse liver
weight.   Repeated daily doses (x7) of a sub-lethal dose
(31.3μg/kg) which produced no apparent increase in liver
weight after one dose was shown to produce an accumulative
effect and resulted in a final liver weight increase of 75%.

Table 2. Mouse response to administration of microcystin-LR
by different routes: organ weight increase

Route	% weight increase: mean (range)*	
	Liver	Kidney
i.p.	50.5 (44.1-56.4)	9.8 (7.8-12.1)
g.i.	43.0 (34.3-50.3)	5.9 (4.8-7.1)
i.n.	41.6 (36.0-48.5)	7.5 (5.7-10.2)

*   6 animals in each group.

Table 3   Microcystin-LR - effect of sub-lethal doses
administered to mice by the intranasal route

Single doses (µg/kg)	% increase in weight
500	87
250	37.4
125	24.4
62.5	1.5
31.3	0

Multiple sub-lethal-low doses
    (x 7-once a day doses)

31.3	75

Histopathology

Intranasal route.   In the nasal mucosa there was
extensive necrosis of the epithelium of both olfactory and
respiratory zones.   The earliest change in olfactory
epithelium was separation of the superficial portions of the
cells, forming blebs containing proteinaceous debris.   This
progressed to destruction of large areas of mucosa to the
level of deep blood vessels.

Liver lesions were consistently present and consisted of
centrilobular necrosis with haemorrhage and accumulation of
large quantities of blood within each lobule.   The earliest
change was vacuolar degeneration and necrosis of groups of
hepatocytes around the central vein. Many of the necrotic/
haemorrhagic zones coalesced with those of adjacent lobules.
There was no evidence of activation of Kupffer cells or any
inflammatory response.

In the adrenal glands there was vacuolation and necrosis
of the inner cortex, particularly the zona reticularis, accom-
panied by intense congestion of the medullary blood vessels.

Lesions were not observed in the trachea, lungs,
oesophagus, pancreas, spleen, lymph nodes, kidneys or brain.

Table 4   Mouse response to administration of anatoxin-a by
           different routes: LD_{50}

           Route $\underline{LD}_{50}$

           i.p.   375 μg/kg
           g.i.   >5000 μg/kg
           i.n.   2000 μg/kg

Table 5   Synergism between Microcystin-LR and Anatoxin-a
           when administered to mice by the intranasal route

       Condition $\underline{LD}_{50}$

       Anatoxin-a only   2000 μg/kg

       Anatoxin-a administered
       30 mins after sub-lethal dose
       (31.3 μg/kg) microcystin-LR   500 μg/kg

    <u>Intraperitoneal and intragastric routes</u>.   Changes in the
liver and adrenal glands identical to those described for the
i.n. route were present.   However, there were no comparable
lesions in the nasal mucosa.

Response to the administration of Anatoxin-a by different routes.

    $\underline{LD}_{50}$. Where deaths occurred they did so within 2 minutes
of anatoxin-a administration and were characterised by loss of
co-ordination, twitching and death by respiratory failure.
Table 4 shows the dose of anatoxin-a required to produce an
LD_{50} when administered by different routes.   The results show
that the most sensitive routes of administration are i.p. at
375 μg/kg followed by i.n. at 2000 μg/kg.   Anatoxin-a administered by g.i. at a much higher dose (5000 μg/kg) failed to
produce any lethalities.

Synergism between Microcystin-LR and Anatoxin-a when administered to mice by the intranasal route.

    $\underline{LD}_{50}$.   At 30 mins prior to administration of anatoxin
i.n., microcystin-LR was given by the same route at a sub-
lethal dose (31.3μg/kg).   The principal effect was to lower
the LD_{50} for anatoxin-a by approximately 4-fold, from 2000
μg/kg to 500 μg/kg (Table 5).

    4   DISCUSSION

Microcystin

    In order to exert hepatotoxicity, microcystin-LR requires
delivery to the liver via the blood stream.   This is achieved
during bioassays by i.p. inoculation of mice[4].   However, this
route cannot be considered relevant to natural exposure via

contaminated water where the most obvious route is oral ingestion, especially as the toxin is resistant to low pH and protein enzymes. This study has shown that although liver damage and death can result from oral ingestion of microcystin-LR, much greater doses are required than when the toxin is administered by the respiratory tract. The oral LD_{50} was 3000 μg/kg, a twelve-fold difference. In this respect the intranasal route is as sensitive as the i.p. route used for laboratory bioassay.

A feature of microcystin-LR administration was the rapid haemorrhage and weight increase which was found to be directly related to dose. Repeated sub-lethal doses showed the effect to be accumulative rising from no weight increase after one dose to an increase of 75% following the seventh daily dose, presumably due to insufficient time for liver repair to be completed. Further evaluation of low dose microcystin levels, frequency and duration would seem to be required.

Although the liver lesions and increase in liver and kidney weight were similar for all routes of administration, the intranasal route presents an additional feature in that necrosis of the nasal mucosa occurs. This would facilitate entry of the toxin into the blood from the nasal vessels and may account in some degree for the greater efficacy of this route when compared to oral ingestion.

In contrast to the intranasal route, inhalation of a fine particle aerosol (3-5 μm in diameter) did not produce any adverse effects in the mouse model. This was undoubtedly because the inhaled dose was extremely low (approximately 0.0005 μg/kg) by virtue of the technique and availability of materials. The technique produces an aerosol of particle size 3-5μm in diameter, which in the main would not experience inertial impaction in the fast moving air stream in the upper, larger airways (unlike the intranasal route) and particles would reach the alveoli, where, in the slow moving air, they would be deposited by gravitational sedimentation. No abnormal histopathology or weight gain was observed with respect to the lung in response to this low dose. However, in the natural environment an aerosol would normally be heterogeneous in particle size with much larger particles (>10μm diameter) predominating. These would be more likely to impact onto mucosal surfaces in the bronchi and nasal areas and would produce an effect akin to administration of the toxin by the intranasal route. This type of aerosol could be produced by waves, water-sports, swimming, animals drinking and algal blooms or scums blown across water surfaces. Additionally, as the toxin tolerates pH conditions ranging from 2 to 4 and is heat-resistant, high concentrations could build up in an aquatic environment. Thus in this respect the aerosol route should be investigated further in terms of microcystin concentration and particle size.

Anatoxin-a: route of administration comparison

As for microcystin-LR, intranasal administration of
anatoxin-a was found to be the most efficient route of
delivery requiring at least two-fold less toxin to produce an
LD_{50} in mice than the oral (g.i.) route.   The efficacy of this
route may result from the direct access of the toxin into the
central nervous system provided by contact with the olfactory
lobe situated within the nasal cavity.

Microcystin-LR and anatoxin-a synergism

Since cyanobacteria can produce a variety of toxic
substances[4] it is likely that the potential host may be
subjected to a combination of toxins.   Cyanobacteria have
also been found to produce neurotoxins[5] which may cause loss of
co-ordination, twitching, irregular breathing and death by
respiratory failure.   If both hepatotoxins and neurotoxins
were present in combination within an aerosol, then inhalation
could strip the mucosal layer and enhance the direct uptake of
the neurotoxin into the brain via the olfactory lobe.   This
hypothesis has been confirmed in the present studies where
prior intranasal administration of microcystin-LR facilitated
the neurotoxic activity of anatoxin-a and thereby lowered the
LD_{50} for that toxin when administered intranasally.

5   CONCLUSION

It has been demonstrated in this study for both microcystin-LR
and anatoxin-a that the intranasal route, equating to large
particle (>10μm diameter) inhalation is the route requiring
least toxin to exert an effect.   Additionally, again by this
route, synergism between these two toxins has been demonstrated.   It is therefore of importance that any assessment
of the risk posed by algal toxins to humans should take into
consideration administration by the inhalation route.

6   REFERENCES

1.   M. Schwimmer, D. Schwimmer. "Algae and Man". Plenum
       Publishing Corporation, New York, 1964, p. 368.

2.   A. Baskerville, R.B. Fitzgeorge, M. Broster, P. Hambleton
       and P.J. Dennis. Lancet, 1982, ii, 1389.

3.   L.J. Reed, and M. Muench. Am. J. Hyg. 1938, 27, 493.

4.   C.W. Keevil. "Public Health Aspects of Cyanobacteria
       (blue-green algae)". Proceedings of a London Seminar.
       Association of Medical Microbiologists, 1991, 91.

5.   W.W. Carmichael, "Natural Toxins: characterisation,
       pharmacology and therapeutics". Pergamon, Oxford, 1989,
       201.

Testing of Toxicity in Cyanobacteria by Cellular Assays

J. E. Eriksson,[1] D. M. Toivola,[2] M. Reinikainen,[2]
C. M. I. Råbergh,[2] and J. A. O. Meriluoto[3]

[1]TURKU CENTER OF BIOTECHNOLOGY, BIOCITY, FIN-20520 TURKU, FINLAND

[2]DEPARTMENT OF BIOLOGY, ÅBO AKADEMI UNIVERSITY, BIOCITY, FIN-20520 TURKU, FINLAND

[3]DEPARTMENT OF BIOCHEMISTRY AND PHARMACY, ÅBO AKADEMI UNIVERSITY, BIOCITY, FIN-20520 TURKU, FINLAND

1 INTRODUCTION

There has been an increasing incidence of reports concerning toxic blue-green algal (cyanobacterial) blooms in the world[1-3]. It is well established that the toxins from these organisms can cause poisoning of wild and domestic animals[1,4,5] and constitute a human health hazard both upon short-term high level exposure as well as upon long-term exposure to low levels of these toxins[4,6-10]. With a global increase in the occurrence of these potentially harmful organisms, the importance of efficient analytical means for screening of toxic blooms has been accentuated. Screening for toxins has been routinely carried out with mouse bioassays. As the regulations concerning toxicity testing with live animals have become more stringent, there has been a growing demand for reliable alternatives to mouse assays. Convenient methods for chemical analysis of microcystins and related liver specific toxins from cyanobacteria have been suggested[11-13]. However, because of the high number of variable toxins[14-18], these methods, which are usually based on different forms of high performance liquid chromatography (HPLC), may not always be infallible since the toxin standard(s) for a particular strain may not be available in a given laboratory and hence the toxicity of some strains may go undetected. Cellular tests based on established cell lines have been suggested[19,20], but since the microcystins and nodularin require an active transport system specific for liver cells[22,23], positive results in cells other than hepatocytes and enterocytes[24] may merely reflect the presence of deleterious compounds in the cyanobacterial extracts other than the actual toxins. In this communication we compare the available options for using cell assays to determine the presence of microcystins in algal extracts.

2 MATERIALS AND METHODS

Isolation of hepatocytes

Hepatocytes were isolated from male Wistar rats (200-250 g) and rainbow trout (*Oncorhynchus mykiss*) by a two-step collagenase perfusion of the liver as previously described[25]. The isolation of fish hepatocytes was modified as described by Råbergh *et al.* 1992[26]. During the perfusion, buffers were supplemented with 0.2-1% bovine serum albumin. All experiments were carried out at 2×10^6 cells/ml in a buffer containing 30 mM Hepes, 30 mM TES, 30 mM Tricine, pH 7.4, 68 mM

NaCl, 5.4 mM KCl, 1.2 mM $CaCl_2$, 0.6 mM $MgCl_2$, 1.1 mM KH_2PO_4, 0.7 mM Na_2SO_4 and 10 mM glucose.

Light and electron microscopy

Hepatocytes were fixed for light microscopy in 3% (w/v) paraformaldehyde in phosphate buffer saline (pH 7.4) and mounted on a glass cover slip before viewing in an Olympus microscope using interference contrast optics. Cells for transmission electron microscopy (TEM) were fixed with 2% (w/v) glutaraldehyde in 0.15 M sodium cacodylate buffer (pH 7.4, RT, 1 h) and postfixed with 1% OsO_4 in 0.15 M sodium cacodylate buffer (pH 7.4, 4°C, 1 h). Samples were dehydrated in a series of ethanol, embedded in Epon resin and stained with lead citrate and uranyl acetate before viewing in a Jeol transmission electron microscope.

Protein phosphatase assays

[32]P-labeled glycogen phosphorylase *a* was obtained by using the commercial kit from Gibco BRL (Gaithesburg, Maryland, USA). Crude cell extracts with type-1 and type-2A phosphatases were obtained by homogenizing a dish (10 cm in diameter) of nearly confluent BHK-21 fibroblasts with 0.5 ml of a buffer containing 20 mM Hepes, pH 7.4, 1 mM $MgCl_2$, 30 mM β-mercaptoethanol, 10% glycerol, 1 mM EGTA, 1 mM PMSF, 0.2% NP-40, 10 µg/ml leupeptin, 10 µg/ml antipain and 1 µg/ml aprotinin on ice. When the cells had been homogenized by passing the extracts 10 times through the tip of a 1 ml micropipettor, the extracts were centrifuged in a table-top centrifuge with 15 000 g for 10 min at 4°C. The supernatants were collected and kept on ice until they were used for measuring the inhibitor potency of algal extracts. The cell extracts were isolated on the same day when the assays were performed.

Cyanobacterial freeze dried material was obtained from natural blooms and from laboratory cultures. 10-20 mg of freeze dried material in 1 ml of water was sonicated with a probe sonicator for 10 s and the resulting homogenate was centrifuged at 15 000 g for 5 min. A dilution series was made from this homogenate with dilutions up to $1:10^{10}$. The inhibitor potency of these extracts was determined by mixing 10 µl of the diluted algal extracts with 10 µl of the crude protein phosphatase preparation (diluted 1:10 with the above-described buffer) and 10 µl of the ^{32}P-labeled phosphorylase *a*. The samples were incubated for 30 min at 30°C, whereafter the reaction was stopped by adding 90 µl 20% TCA. Samples were kept on ice for 10 min, centrifuged at 15 000 g for 3 min and the supernatant was collected. The protein phosphatase activity in the samples was determined as released radioactivity in the supernatants, as measured by scintillation counting. Microcystin-LR (MC-LR) was isolated from a bloom material of *Microcystis aeruginosa* as previously described[12,27,28].

2 RESULTS AND DISCUSSION

One of the most obvious approaches for in vitro testing of any kind of toxic compound is to use cultures of established cell lines. It has been suggested in a couple of previous studies that this approach could be useful also for microcystins and related toxins[19,20,29]. However, it has been shown in many different studies that only parenchymal liver cells respond to these toxins at concentrations analogous to the doses effective in the whole animal[30,31]. This cellular specificity seems to be due to a cell selective uptake mechanism, the multispecific bile acid transport system, as first suggested by Runnegar *et al.* 1981[21]. The cellular uptake specificity was confirmed in a previous study where we studied the uptake of a tritiated derivative of microcystin-LR, ^{3}H-dihydromicrocystin-LR, in three different established cell lines and compared it to the uptake in freshly isolated liver cells[22]. The study included a fibroblast cell line (NIH-3T3), a neuroblastoma cell line (SH-SY5Y), and a liver cell line (Hep-G2). We observed that the uptake in the established cell lines was negligible compared to the uptake in hepatocytes, the latter reaching rapidly a plateau already after 10-20 min (Figure 1). It was also shown that the uptake could be competitively inhibited by excess amount of bile acids[22]. These results imply that the obvious cellular specificity of microcystins is due to the fact that the intracellular concentrations of these toxins can reach effective levels only in hepatocytes which posses the required transport mechanism. Interestingly, however, upon long exposure (10-24 h) of some cultured cell lines, we could observe some uptake of the labelled microcystin (Figure 1). This is probably due to pinocytotic activity. This could explain why very high concentrations of microcystins (> 100 μM) can produce morphological effects in cells other than hepatocytes when incubated for sufficiently long periods (>10 h; Wikstrom, M. *et al.* unpublished results). The uptake of small amounts of the toxins may also have ramifications in terms of the suggested tumor-promoting effect of microcystins[9,32,22], since very small amounts may be sufficient to

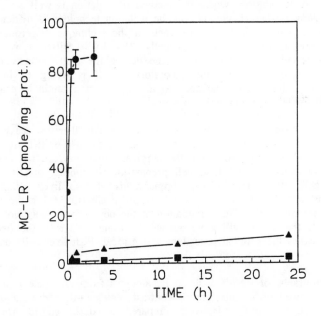

Figure 1 Time-dependent uptake of [3]HDMC-LR in isolated rat hepatocytes (●), 3T3-fibroblasts (▲) and a neuroblastoma cell line (SH-SY5Y) (■). The uptake is presented as pmole MC-LR/mg cell protein. Notice the uptake is very rapid in hepatocytes and negligible in the other cell types at early time points. At 24 h the established cell lines show some uptake, probably due to pinocytosis. Data from Eriksson *et al.* 1990[22] with permission.

induce tumor promotion although no major effects can be observed morphologically in the cells. However, for the purpose of toxicity testing, established cell lines are not useful since they do not express the transport proteins required for bile acid uptake. It is also documented that even primary liver cell cultures cease to express the transport proteins when maintained in culture for more than 2-3 days[34]. Positive responses obtained in cells other than freshly isolated hepatocytes are thus likely to reflect the presence of harmful compounds in the algal extracts other than the actual toxins.

In freshly isolated rat liver cells the morphological effects of microcystins are very rapid and obvious. They can easily be observed by using ordinary phase contrast light microscopy or interference contrast light microscopy (Figure 2 a and c). These toxins cause a hepatocyte deformation which is rather specific for these compounds. The characteristically clustered blebs are very conspicuous in electron microscopy (Figure 2 e and g) but can also be distinguished in the light microscope. Thus, it should be relatively easy to distinguish toxin-specific effects even in the case where there were some unspecific harmful compounds in the algal extracts, since the morphological effects they would produce would be different from those of microcystins. Rat liver cells have been successfully used to screen for toxic cyanobacterial blooms[35]. In this particular study the authors could show a very good correlation between the effects on liver cells and the toxicity in mouse assays. On the whole, it appears that freshly isolated rat liver cells could be a useful model for toxicity testing in laboratories where the method of isolation is well established. The drawback of using rat hepatocytes is that the isolation technique is difficult and only laboratories with a great deal of experience in the technique can routinely obtain high quality preparations of rat liver cells. Rat liver cells, as well as other mammalian liver cells, are prone to damage during the isolation procedure and it will require a major effort of any laboratory which is trying to initialize this technique before it has been established to a degree which routinely produces cells with a viability of 90% or more.

Laboratories operating in an environment where there is easy access to laboratory or farm raised rainbow trout, may consider the use of fish liver cells as a substitute for rat liver cells. Although the surgical techniques are not easier in fish compared to those used in rat liver cell preparation, it appears that it is relatively easy to routinely obtain high quality preparations of fish liver cells. There are several reasons why fish liver cells may be more easily isolated without damaging the cells than rat liver cells. The cannulation of the hepatic vein does not have to be performed while the fish is still alive but can be done soon after the fish has been killed, which makes the surgical procedure less critical. Fish liver cells have a much

Figure 2 Micrographs of freshly isolated hepatocytes from rat (a,c,e,g) and rainbow trout (b,d,f,h) viewed under interference contrast optics (a-d) and in a transmission electron microscope (e-h). Cells were prepared as described in Materials and methods. Control cells are shown in a,b,e and f. Rat hepatocytes were treated with 1 μM MC-LR for 60 minutes, (c,g) and fish hepatocytes with 10 μM MC-LR for 60 minutes (d, h). Bars: a, 30 μm (applies for a-d); e, 5 μm (applies for e and g); f, 5 μm (applies for f and h). Fish hepatocytes are smaller than rat hepatocytes and have a very high content of glycogen granules (white vacuoles). Notice the resemblance of the microcystin-induced effects in both rat and fish hepatocytes.

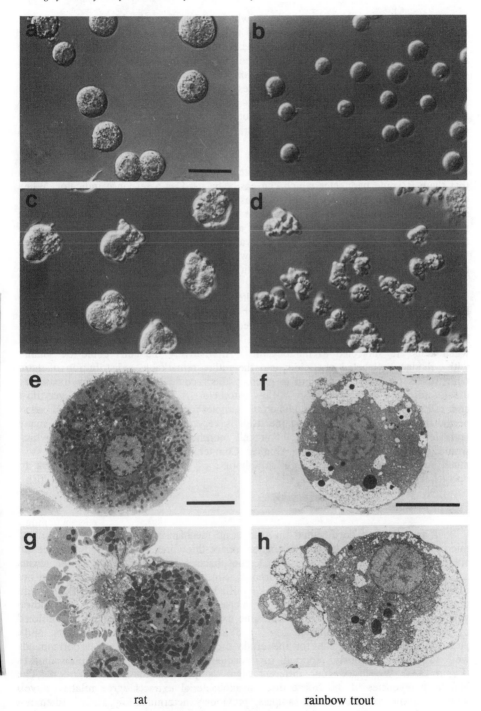

rat   rainbow trout

<u>Figure 2</u>   (See bottom of page 78 for legend)

lower metabolic rate than rat liver cells which makes the cells less prone to anoxic damage. If the fish have been maintained at 18-20 °C, cell incubations can be done at room temperature which also makes these cells well suited for toxicity tests since no water baths or incubators are required. Evidently fish liver cells respond very well to microcystins. The morphology of the cells is affected approximately at concentrations corresponding to the doses effective *in vivo* in fish[36] (Figure 2). Interestingly, the morphological alterations resemble those produced in rat liver cells (Figure 2). Also in this case the alterations are easy to observe under normal light microscope. Fish liver cells are, however, somewhat less sensitive to microcystins than rat liver cells.

One precondition for any cell type to be used for routine bioassay purposes is that there is some parameter which can easily be used as a response indicator. When toxicity is measured *in vitro*, the response is ideally monitored as a biochemical response. If the viability of the cells to be used for the bioassay is affected, then the response can be conveniently monitored as lactate dehydrogenase leakage or with any other of the numerous options available for measuring alterations in cell viability. With respect to microcystins, however, this possibility is not available since these toxins do not rapidly alter the viability of the cells[30].

The observation of morphological alterations by microscopy is easy to perform but this is a time-consuming parameter to monitor in routine toxicity testing. Alterations in morphology can be monitored mechanically by using a flow-cytometer. In a previous study it was shown that the forward-angle light scatter signal is altered when cells are subjected to microcystin[37]. Alternatively the cells could be stained with rhodamine-labeled phalloidin and the rather remarkable redistribution of actin which takes place upon exposure to microcystins[38] could be monitored by measuring the fluorescence distribution with a flow-cytometer. A flow-cytometer is a very fast and effective device by which numerous samples can be processed within a short period of time. However, the machine itself is very costly and not available in many laboratories. Possibly the alterations in cell morphology are of a magnitude which would enable monitoring with a Coulter Counter. This cell counting device is a much more abundant tool in routine laboratories and is far less complicated to operate than a flow-cytometer.

Protein phosphatase inhibition

It is now well documented that the basic mechanism underlying the action of microcystins is inhibition of the major serine-threonine protein phosphatases in eukaryotic cells, i.e. type-1 and type-2A phosphatases[39-44]. This feature of the toxins can readily be used for monitoring purposes. Since pure protein phosphatases are not easily available we explored a highly simplified approach for employing the protein phosphatases to detect the presence of microcystins. We used a crude preparation which is basically comprised of cytoplasmic extracts of BHK-21 fibroblasts stabilized by the presence of glycerol and Mg^{2+} and a cocktail of protease inhibitors. With pure MC-LR the IC_{50} value for the crude protein phosphatase extract corresponds well to those previously reported for purified type-1 and type-2A phosphatases (Figure 3A). We tested four different algal samples on the crude extracts. The inhibitory potencies of the freeze dried cyanobacterial extracts agree relatively well with the toxin contents of the samples, previously determined by HPLC (data not shown). If a pure standard with a known concentration is used in parallel each time a new protein phosphatase preparation is made, it should be possible to calculate a

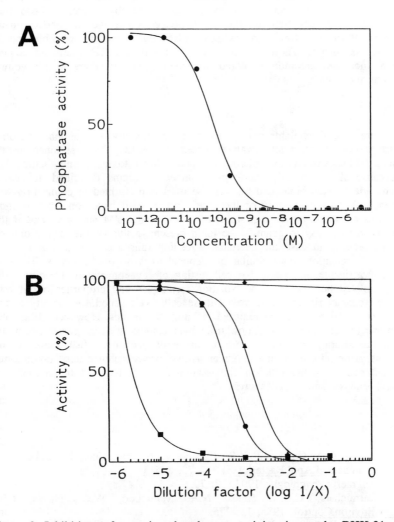

<u>Figure 3</u> Inhibition of protein phosphatase activity in crude BHK-21 fibroblast extracts isolated as described in materials and methods. (A) Concentration dependent inhibition of protein phosphatase activity by purified MC-LR. (B) Inhibition of protein phosphatase activity by dilutions of crude extracts of cyanobacterial blooms and cultures: (■) a natural bloom of *M. aeruginosa* (MC-LR 6.7 µg/mg dry weight), (●) a laboratory culture of *M. aeruginosa* (MC-LR 2.0 µg/mg dry weight), (▲) a natural bloom of *Oscillatoria agardhii* (MC-RR, 1.9 µg/mg dry weight) and (◆) a non-toxic natural bloom of *M. aeruginosa* (no toxin peak detected on HPLC). HPLC analysis was performed as previously described[12].

relatively exact toxin concentration in the samples. Theoretically it should be possible to calculate the toxin concentration in an unknown algal sample by using the IC_{50}-value of the known standard, since the IC_{50}-value in the dilution series should contain exactly the same amount of toxin. In our small number of samples we could get a relatively good but not perfect correlation between the toxin content measured by HPLC and the concentration estimated by protein phosphatase inhibition. The discrepancy between the actual concentration and the one estimated by the protein phosphatase inhibition assay could be due to several possible sources of errors in the assays. These sources of errors will not be examined in this presentation since they are more explored in greater detail elsewhere in this volume (see also Sim and Mudge 1993[45]).

Conclusions

There are several available options for assessing toxicity of cyanobacteria other than mouse assays or chemical analysis based on HPLC. As mentioned in the Introduction, HPLC analysis of microcystin and related toxins is not satisfyingly reliable because of the rapidly increasing number of toxins described related to microcystins. Ideally, the biological assay to be used is performed in connection with chemical analysis. In the majority of the cases, blooms still have some of the more common forms of microcystins (e.g. -LA, -LR, -RR, -YR, -YM variants), and if this is the case it will be easy to quantify the toxin present in the sample. The biological test provides security in case the chemical screening fails and should also rule out false peaks with retention times similar or identical to those of the toxins. The only reliable and practicable alternative for cell testing of cyanobacterial toxicity is to use freshly isolated hepatocytes. The isolation techniques and properties of rat liver cells are well characterized, which makes these cells well suited for toxicity testing in laboratories experienced in the required isolation techniques. However, these cells are very prone to cell damage. Rainbow trout liver cells may be an easier alternative for novice laboratories, provided there is an easy access to fish close to the concerned laboratory. In addition to cellular assays, protein phosphatase inhibition is a simple and rather reliable biochemical assay which can be used in parallel with the chemical analysis and/or cell assays.

REFERENCES

1. O.M. Skulberg, G.A. Codd and W.W. Carmichael, Ambio, 1984, 13, 244.
2. W.W. Carmichael, J.Appl.Bacteriol., 1992, 72, 445.
3. G.M. Hallegraeff, Phycologia, 1993, 32, 79.
4. W.W. Carmichael, C.L.A. Jones, N.A. Mahmood, W.C. Theiss. CRC Crit.Rev.Environ.Control, 1985, 15, 275.
5. C. Edwards, K.A. Beattie, C.M. Scrimgeour and G.A. Codd, Toxicon, 1992, 30, 1165.
6. I.R. Falconer, A.M. Beresford, M.T.C. Runnegar, Med.J.Aust., 1983, 1, 511.
7. P.R. Hawkins, M.T.C. Runnegar, A.R.B. Jackson and I.R. Falconer, Appl.Environ.Microbiol., 1985, 50, 1292.
8. G.A. Codd and G.K. Poon, 'Biochemistry of the algae', (eds. L.J. Rogers and J.R. Gallon), Oxford Science Publ., Clarendon Press, Oxford. 1988, Vol. 28, p. 283.
9. I.R. Falconer, Environ.Toxicol.Water.Quality, 1991, 6, 177.
10. H. Fujiki, Mol.Carcinogen., 1992, 5, 91.
11. G.K. Poon, I.M. Priestley, S.M. Hunt, J.K. Fawell and G.A. Codd, J.Chromatogr., 1987, 387, 531.
12. J.A.O. Meriluoto and J.E. Eriksson, J.Chromatogr., 1988, 438, 93.

13. K.-I. Harada, K. Matsuura, M. Suzuki, H. Oka, M.F. Watanabe, S. Oishi, A.M. Dahlem, V.R. Beasley and W.W. Carmichael, J.Chromatogr., 1988, 448, 275.
14. K. Sivonen, M. Namikoshi, W.R. Evans, W.W. Carmichael, F. Sun, L. Rouhiainen, R. Luukkainen and K.L. Rinehart, Appl.Environ.Microbiol., 1992, 58, 2495.
15. K. Sivonen, O.M. Skulberg, M. Namikoshi, W.R. Evans, W.W. Carmichael and K.L. Rinehart, Toxicon, 1992, 30, 1465.
16. T. Kondo, Y. Ikai, H. Oka, N. Ishikawa, M.F. Watanabe, M. Watanabe, K.-I. Harada and M. Suzuki, Toxicon, 1992, 30, 227.
17. R. Luukkainen, K. Sivonen, M. Namikoshi, M. Färdig, K.L. Rinehart and S.I. Niemelä, Appl.Environ.Microbiol., 1993, 59, 2204.
18. M. Namikoshi, B.W. Choi, F.R. Sun, K.L Rinehart, W.R. Evans and W.W. Carmichael, Chem.Res.Toxicol., 1993, 6, 151.
19. W.W. Carmichael and P.E. Bent, Appl.Environ.Microbiol., 1981, 41, 1383.
20. W.O.K. Grabow, W.C. Du Randt, O.W. Prozesky and W.E. Scott, Appl.Environ.Microbiol., 1982, 43, 1425.
21. M.T. Runnegar, I.R. Falconer and J. Silver, Naunyn-Schmideberg's Arch.Pharmacol., 1981, 317, 268.
22. J.E. Eriksson, L. Grönberg, S. Nygård, J.P. Slotte and J.A.O. Meriluoto, Biochim.Biophys.Acta, 1990, 1025, 60.
23. M.I.T. Runnegar, R.G. Gerdes and I.R. Falconer, Toxicon, 1990, 29, 43.
24. I.R. Falconer, M. Dornbusch, G. Moran and S.K. Yeung, Toxicon, 1992, 30, 790.
25. P.O. Seglen, 'Methods in Cell Biology', (ed. D.M. Prescott), Academic Press, New York, 1976, Vol XIII, p. 29.
26. C.M.I. Råbergh, B. Isomaa and J.E. Eriksson, Aquatic Toxicol., 1992, 23, 169.
27. J.E. Eriksson, J.A.O. Meriluoto, H.P. Kujari and O.M. Skulberg, Comp.Biochem.Physiol., 1988, 89C, 207.
28. J.E. Eriksson, J.A.O. Meriluoto, H.P. Kujari, K. Österlund, K. Fagerlund and L. Hällbom, Toxicon, 1988, 26, 161.
29. Y.A. Kirpenko, V.V. Stankevich, V.M. Orlovskiy, N.I. Kirpenko, A.V. Bokov and T.F. Karpenko, Hydrobiol.J., 1980, 15, 83.
30. J.E. Eriksson, H. Hägerstrand and B. Isomaa, Biochim.Biophys.Acta, 1987, 930, 304.
31. I.R. Falconer and M.T.C. Runnegar, Chem.-Biol.Interactions, 1987, 63, 215.
32. I.R. Falconer, Med.J.Austr., 1989, 150, 351.
33. R. Nishiwaki-Matusushima, T. Ohta, S. Nishiwaki, M. Suganuma, K. Kohyama, T. Ishikawa, W.W. Carmichael and H. Fujiki, J.Cancer Res.Clin.Oncol., 1992, 118, 420.
34. W. Fällman, E. Petzinger and R.K.H. Kinne, Am.J.Physiol., 1990, 258, C700.
35. T. Aune and K. Berg, J.Toxicol.Environ.Health, 1986, 19, 325.
36. C.M.I. Råbergh, G. Bylund and J.E. Eriksson, Aquatic Toxicol., 1991, 20, 131.
37. J.E. Eriksson, J.A.O. Meriluoto, H.P. Kujari, K. Jamel Al-Layl and G.A. Codd, Toxicity Assess., 1988, 3, 511.
38. J.E. Eriksson, G.I.L. Paatero, J.A.O. Meriluoto, G.A. Codd, G.E.N. Kass, P. Nicotera and S. Orrenius, Exp.Cell Res., 1989, 185, 86.
39. R.E. Honkanen, J. Zwillers, R.E. Moore, S.L. Daily, B.S. Khatra, M. Dukelow and A.L. Boynton, J.Biol.Chem., 1990, 265, 19401.
40. C. MacKintosh, K.A. Beattie, S. Klumpp, P. Cohen and G.A. Codd., FEBS Lett., 1990, 264, 187.
41. J.E. Eriksson, D. Toivola, J.A.O. Meriluoto, H. Karaki, Y.-G. Han and D. Hartshorne, Biochem.Biophys.Res.Commun., 1990, 173, 1347.

42. S. Yoshizawa, R. Matsushima, M.F. Watanabe, K.-I. Harada, A. Ichihara, W.W. Carmichael and H. Fujiki, J.Cancer Res.Clin.Oncol., 1990, 116, 609.
43. R. Matsushima, S. Yoshizawa, M.F. Watanabe, K.-I. Harada, M. Furusawa, W.W. Carmichael and H. Fujiki, Biochem.Biophys.Res.Coummun., 1990, 171, 867.
44. R.E. Honkanen, M. Dukelow, J. Zwiller, R.E. Moore, B.S. Khatra and A.L. Boynton, Mol.Pharmacol., 1991, 40, 577.
45. A.T.R. Sim and L.-M. Mudge, Toxicon, 1993, 31, 1179.

A Sensitive Bioscreen for Detection of Cyclic Peptide Toxins of the Microcystin Class

Charles F. B. Holmes,[1] Tara L. McCready,[1] Marcia Craig,[1] Timothy W. Lambert,[2] and Steve E. Hrudey[2]

[1]MRC CANADA PROTEIN STRUCTURE AND FUNCTION GROUP, DEPARTMENT OF BIOCHEMISTRY, UNIVERSITY OF ALBERTA, EDMONTON TG 2H7, CANADA

[2]DEPARTMENT OF PUBLIC HEALTH SCIENCES, UNIVERSITY OF ALBERTA, EDMONTON TG 2H7, CANADA

1 ABSTRACT

We have developed a quantitative bioscreen which will detect, with unprecedented sensitivity, hepatotoxic cyclic peptides of the microcystin class in marine and freshwater organisms. Our method employs capillary electrophoresis (CE) coupled with liquid chromatography (LC)-linked protein phosphatase (PPase) enzyme bioassay. The detection protocol therefore combines the precision of instrumental analytical techniques with direct quantitation of the actual biological activity of toxins present. Since the microcystins are potent inhibitors of the catalytic subunits of eukaryotic type-1 and -2A protein phosphatases (PP-1c and PP-2Ac), crude extracts are first quantitatively assayed for their ability to inhibit PP-1c and/or PP-2Ac. Following detection, toxins are fractionated by a PPase assay-guided two-step LC protocol at pH 6 and at pH 2. This procedure provides a detection limit of 1-5 pg toxin. When levels present exceed 1 ng, CE is applied to rapidly isolate PPase inhibitors partially purified by reverse phase LC, thus enabling optical detection of the microcystins at 200 nm. The unified LC/CE-linked PPase bioassay facilitates resolution of multiple structural variants of the microcystins, is sufficiently sensitive to detect pg quantities of microcystins in drinking water and should assist in establishing acceptable quarantine levels for human consumption of these toxins.

2 INTRODUCTION

Cyanobacterial hepatotoxins of the heptapeptide microcystin and pentapeptide nodularin classes represent a health threat to world-wide drinking water supplies, providing dangerous implications for both human and agricultural livestock consumption (reviewed extensively by Carmichael [1] and references therein). In addition, we recently provided evidence for the presence of these toxins in the marine environment.[2-4] Microcystins/nodularins are potent, specific inhibitors of the catalytic subunits of protein phosphatase-1 and -2A (PP-1c/2Ac), two of the major serine/threonine protein phosphatases involved in eukaryotic cell regulation.[5-8] We have exploited these properties to develop a sensitive bioscreen for their detection based on protein phosphatase (PPase) bioassay linked to instrumental analytical techniques.

3   EXPERIMENTAL PROCEDURES

Cyanobacteria (containing predominantly *Microcystis aeruginosa*) were collected from a bloom present on Little Beaver Lake, Alberta, Canada (the drinking water supply for the town of Ferintosh, Alberta) during August 1991. The cyanobacteria were recovered by centrifugation (4,000 g for 30 min) and lyophilized. Portions (30 g) of this material were extracted four times by disruptive Polytron homogenization in methanol. Aliquots (1 μl) from the extract were analysed for their ability to inhibit the dephosphorylation of ^{32}P-radiolabelled glycogen phosphorylase *a* by PP-1c purified from rabbit skeletal muscle. Further details of this procedure were fully described previously.[9,10]

For preparative isolation of cyanobacterial toxins, active supernatants from methanolic extraction of lyophilised cyanobacteria were extracted with 8 vol. of hexane, and concentrated on a Speedvac concentrator. Samples were fractionated on a Sephadex LH-20 chromatography column (20 mm x 900 mm), and eluted in methanol with a flow rate of 0.25 ml/min. Eluent from this column separation was assayed for PP-1c inhibitory activity and the active fractions pooled and dried. These fractions were fractionated by a PP-1c bioassay-guided two-step reverse phase C_{18} LC protocol at pH 6.5 and pH 2.0. Purified microcystins were analysed by amino acid analysis and mass spectrometry.[11] Capillary electrophoresis (CE) of microcystins was carried out on a Beckman 2100 PACE instrument in 10 mM Tris, pH 6.0, with a capillary of 75 μm internal diameter x 50 cm effective length, applied voltage of 20 kV and detection at 200 nm.

4   RESULTS

Identification and Characterisation of Novel Microcystins in Cyanobacteria from an Alberta Drinking Water Lake

Application of a novel two-step reverse LC protocol allowed for the effective detection and isolation of several microcystin analogues from cyanobacteria collected from an Alberta drinking water lake. The microcystins identified included microcystin-LR, -FR, -LA and XR (where X is a novel amino acid of molecular mass 193).[10] In addition, several novel microcystins were identified from the same cyanobacteria which had the variable arginine residue in the cyclic heptapeptide replaced by a variety of hydrophobic amino acids (Table 1). These relatively hydrophobic microcystins were purified to homogeneity and inhibited PP-1c catalysed dephosphorylation of ^{32}P-radiolabelled phosphorylase *a* with a potency similar to that of microcystin-LR (*i.e.* their IC_{50} values were in the expected range of 0.1 - 0.4 nM).

Table 1   *Structures of novel hydrophobic cyclic peptide microcystins*

Peptide	Structure	IC_{50} [nM] vs PP-1c
Microcystin-LV	cyclo(D-Ala-L-Leu-D-McAsp-L-Val-Adda-D-Glu-Mdha)	0.3
Microcystin-LM	cyclo(D-Ala-L-Leu-D-McAsp-L-Met-Adda-D-Glu-Mdha)	0.1
Microcystin-LL	cyclo(D-Ala-L-Leu-D-McAsp-L-Leu-Adda-D-Glu-Mdha)	0.1
Microcystin-LF	cyclo(D-Ala-L-Leu-D-McAsp-L-Phe-Adda-D-Glu-Mdha)	0.4

Identification and Quantitation of Microcystins in a Canadian Drinking Water Supply

The LC-PPase bioassay procedure was directly applied to the analysis of microcystins in raw (R1-11) and treated (Tap 1-7) drinking water from Little Beaver Lake, Alberta during August 1992. The results show a sample collection carried out on one day between 9:00 and 20:30 hours. Following detection by PP-1c inhibition assay, the predominant microcystin was identified as microcystin-LR by two-step LC analysis linked to the PPase bioassay. Typical levels of microcystin-LR in samples of tap water supplied by this lake varied between 0.1-1.0 µg/L (Table 2).

Table 2   *Microcystin-LR levels in raw and treated drinking water*

Water Sample	Time	Toxin	Std Deviation (µg/L Microcystin-LR)	Water Sample	Time	Toxin	Std Deviation (µg/L Microcystin-LR)
R 1	09:00	1.89	0.23				
R 2	10:00	0.54	0.03				
R 3	11:00	0.83	0.06				
R 4	12:00	0.52	0.11	Tap 1	12:00	0.38	0.05
R 5	13:00	0.57	0.01	Tap 2	13:00	0.28	0.07
R 6	14:20	0.97	0.01	Tap 3	14:20	0.37	0.04
R 7	15:00	1.00	0.02	Tap 4	15:00	0.32	0.09
R 8	17:30	0.45	0.03	Tap 5	17:30	0.31	0.08
R 9	18:30	2.41	0.30	Tap 6	18:30	0.31	0.09
R 10	19:30	1.83	0.22	Tap 7	20:30	0.37	0.06
R 11	20:30	1.69	0.21				

5   DISCUSSION

The LC-PPase bioassay has been successfully applied to the identification and isolation of a wide variety of microcystins in cyanobacteria isolated from a Canadian drinking water lake. In addition, it has been used to detect these compounds in raw and treated tap water supplied by this lake. The detection sensitivity of these procedures for the analysis of microcystins is approximately 2 pg per µl of extract. This allows the facile detection of microcystins in drinking water supplies at a level of 2 µg per litre without sample concentration. Since it is relatively easy to concentrate water contaminated with microcystins by at least 100-fold, PPase bioassay alone should have a useful practical application for detection of microcystins at levels from 0.02-2 µg per litre of water. In practice, when the presence of microcystins is suspected from initial PPase bioassay, it is possible to further identify the nature of microcystins involved by utilising a two-step reverse phase LC protocol linked to PPase inhibition assay. Virtually unambiguous detection of microcystins in the extract is provided by the powerful resolution capabilities of capillary electrophoresis which may be applied when ng levels of microcystin are present.

The isolation of several new microcystins in this study also indicates that an important future property of the LC/CE PPase bioscreen will be the demonstrated ability of the procedure to identify novel natural toxins with potent biological activity from a variety of marine and freshwater samples. A scheme for potential utilisation of this bioscreen is shown in Figure 1.

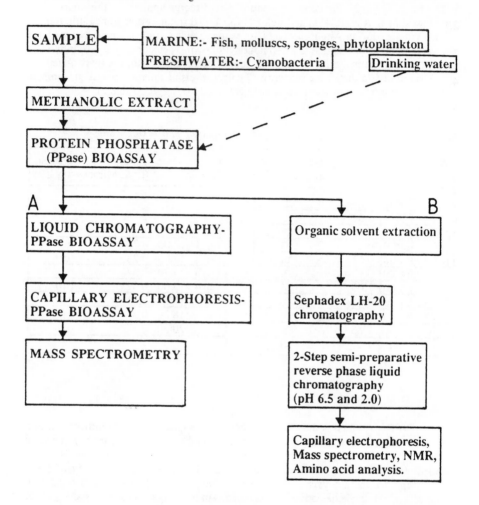

Figure 1   A scheme outlining the identification and characterisation of protein phosphatase inhibitors from marine and freshwater sources (Slightly modified from Ref. 6). Route A applies to analytical scale detection. Route B applies to preparative scale isolation.

6   ACKNOWLEDGEMENTS

We would like to thank Dr R.J. Andersen, University of British Columbia, Vancouver, Canada for valuable discussion. This research was funded by an MRC and NSERC Canada strategic grant to CFBH, and an NSERC strategic grant to SEH.

REFERENCES

1.   W.W. Carmichael, J. Applied Bacteriol., 1992, 72, 445.
2.   S. D. DeSilva, D.E. Williams, R.J. Andersen, H. Klix, C.F.B. Holmes and T.M. Allen, Tetrahedron Lett., 1992, 33, 1561.
3.   R.J. Andersen, H.A. Luu, D.Z.X. Chen, C.F.B. Holmes, M.J. Kent, M. LeBlanc, F.J.R. Taylor and D.E. Williams, Toxicon, 1993, 31, 00.
4.   D.Z.X. Chen, M.P. Boland, C. Ptak, M.A. Smillie, H. Klix, R.J. Andersen and C.F.B. Holmes, Toxicon, 1993, in press.
5.   R. E. Honkanen, J. Zwiller, R.E. Moore, S.L. Daly, B.S. Khatra, M. Dukelow and A.L. Boynton, J. Biol. Chem., 1990, 265, 19401.
6.   S. Yoshizawa, R. Matsushima, M.F. Watanabe, K.I. Harada, K. Ichihara, W.W. Carmichael and H. Fujiki, J. Cancer Res. Clin. Oncol., 1990, 116, 609.
7.   C. Mackintosh, K.A. Beattie, S. Klumpp, P. Cohen and G.A. Codd, FEBS Lett., 1990, 264, 187.
8.   M.J. Hubbard and P. Cohen, Trends Biochem Sci., 1993, 18, 172.
9.   C.F.B. Holmes, Toxicon, 1991, 29, 469.
10. M.P. Boland, M.A. Smillie, D.Z.X. Chen and C.F.B. Holmes, Toxicon, 1993, in press.
11. M. Craig, T.L. McCready, H.A. Luu, M.A. Smillie, P. Dubord and C.F.B. Holmes, Toxicon, 1993, in press.

The Inhibition of Protein Phosphatases by Toxins: Implications for Health and an Extremely Sensitive and Rapid Bioassay for Toxin Detection

Carol MacKintosh and Robert W. MacKintosh

MEDICAL RESEARCH COUNCIL PROTEIN PHOSPHORYLATION UNIT,
DEPARTMENT OF BIOCHEMISTRY, UNIVERSITY OF DUNDEE, DUNDEE
DD1 4HN, UK

SUMMARY

The cyanobacterial microcystins exert their toxic effects because they are extremely potent and specific inhibitors of protein phosphatases 1 and 2A, two classes of enzymes that act as 'molecular control switches' and regulate many processes (cell division and growth, metabolism, hormonal control and so on) inside animal and plant cells. A major interest of our laboratory is to use the microcystins as pharmacological tools to investigate the regulation of cellular processes ranging from human cell differentiation to plant defence responses to fungal attack. However, an important spin-off of this research is the realisation that microcystin can be detected and quantified in drinking water supplies by testing for inhibition of protein phosphatases 1 and 2A in a simple and quick bioassay.

1. PROTEIN PHOSPHATASES 1 AND 2A; TWO CLASSES OF 'MOLECULAR SWITCH' IN ANIMAL AND PLANT CELLS

Every animal and plant cell is able to detect and respond to signals from its environment. For example, heart muscle cells exposed to the hormone adrenalin will contract and relax with increased rate and strength (a feeling familiar to nervous conference speakers!); a *Paramecium* will swim towards chemical attractants and away from repellents; leaves of Mimosa plants move when the surface cells detect touch; and embryonic cells grow and divide in response to growth hormones.

As more-and-more new signals are discovered and their functions are deciphered, a unifying general theory of **signal transduction** is beginning to emerge which encompasses the mechanisms that control all of the specialised responses of plant and animal cells. In outline, each signal is recognised specifically by a different receptor protein, usually on the surface of the cell. When activated, the receptors pass on a chemical message to internal proteins whose role is to amplify and co-ordinate different incoming signals. Finally, the message reaches enzymes which act as 'molecular switches' to trigger the cell's response (Figure 1).

<u>Figure 1</u> shows how the "molecular switches" operate:-

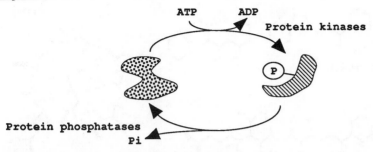

Enzymes called **protein kinases** transfer phosphate groups from adenosine-5'-triphosphate (ATP) to specific serine, threonine or tyrosine residues in a **target protein**. This causes the target protein to change its shape and function. Enzymes called **protein phosphatases** remove the phosphate group and the target protein reverts to its original state. There are four different types of protein phosphatase that act on phosphorylated serine and threonine residues in target proteins and these phosphatases are classified as protein phosphatases 1, 2A, 2B and 2C (PP1, PP2A, PP2B and PP2C).[1]

There are thousands of potential target proteins that are controlled by reversible phosphorylation in cells, including metabolic enzymes, ion channels, components of the cell division apparatus and structural and contractile proteins. For example, phosphorylation of a smooth muscle protein leads to muscle contraction and dephosphorylation to relaxation. The control of contraction and relaxation of smooth muscle in the capillary walls determines blood pressure.

2.   MICROCYSTINS ARE POTENT INHIBITORS OF PP1 AND PP2A
     FROM ANIMALS, FUNGI AND PLANTS

Luckily, the cell signalling systems which regulate the kinases and phosphatases do not go wrong very often. But when cells in animal tissues do lose normal control and respond inappropriately to signals, the result is often a disease like cancer, diabetes or an immunological disorder. So imagine the chaos that would result if a toxin entered a cell and completely blocked the function of one set of molecular control switches. This is exactly how the cyanobacterial toxins **microcystin**[2] and **nodularin**[3] act on PP1 and PP2A!

In addition to the **microcystins** and **nodularins**, PP1 and PP2A are the targets of several chemically-diverse toxins (Figure 2). These protein phosphatase inhibiting compounds include the polyketides **okadaic acid**[4,5] **and related compounds** which are produced by marine dinoflagellates and are a major cause of diarrhetic shellfish poisoning; **calyculin A,**[6] another marine toxin; **tautomycin**[7] from species of the soil bacterium *Streptomyces*; **cantharidin,**[8,9] a terpenoid from blister beetles and otherwise known as the notorious (but reputably disappointing!) aphrodisiac, Spanish Fly; and **endothall,**[8,9] a chemically synthesised herbicide used to clear waterways and to defoliate cotton crops.

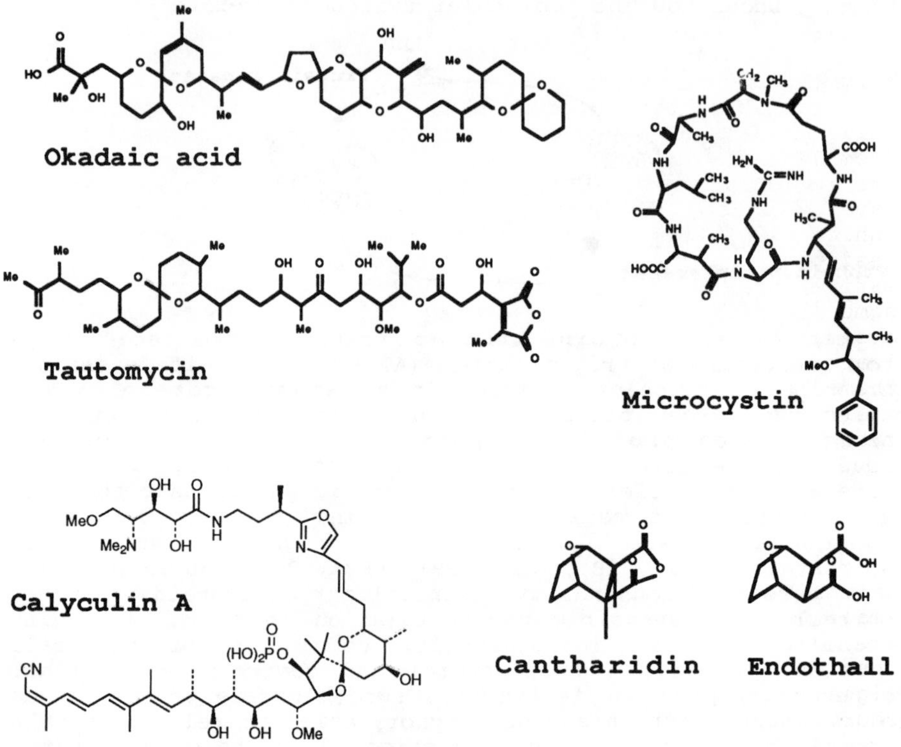

Okadaic acid

Tautomycin

Microcystin

Calyculin A

Cantharidin   **Endothall**

Figure 2   Protein phosphatases PP1 and PP2A are the targets of several toxins from diverse sources

How do we explain the fact that the predominating symptoms caused by microcystins (liver disease), okadaic acid (diarrhetic poisoning) and cantharidin (skin irritation) are all different if inhibition of PP1 and PP2A is the underlying cause of toxicity of all of these compounds? There are two likely reasons. First, toxins show different permeabilities. For example, microcystin gets into liver cells easily.[10] Second, each toxin has a different overall and relative potency against enzymes in the PP1 and PP2A classes.[2-9]

It seems remarkable that such a chemically-diverse group of secondary metabolites derived from freshwater, marine and terrestrial species of prokaryotes, dinoflagellates and insects all bind to PP1 and PP2A and perhaps (as kinetic and binding data suggests[2,3,6,7,8]) even to the same site on these enzymes. Is it reasonable to propose that the protein phosphatase inhibitors may have evolved as defence and attack compounds? Consideration of the evolutionary history of the PP1 and PP2A enzymes makes this idea attractive: Molecular genetic analysis has revealed that the plant, animal and fungal versions of PP1 and PP2A are remarkably similar, with amino acid sequence identities of 70-80% across all eukaryotic phyla.[11] This means that PP1 and PP2A have been present in a huge variety of organisms, living in diverse ecological niches, since before the divergence of plants, animals, and fungi - at least 1000 million years. So perhaps, it is not too surprising if chemical methods for inhibiting these enzymes have arisen several times during evolution.

Available evidence suggests that the protein phosphatases in cyanobacteria (and other prokaryotes) are quite different from PP1 and PP2A,[2] which would explain why these organisms are resistant to poisoning. Some eukaryotic species appear to have evolved mechanisms to counteract the effects of inhibitors, for example *Paramecium* has a PP2A which is resistant to inhibition by okadaic acid[12] and microcystin;[2] marine mussels store okadaic acid in their hepatopancreas; and some zooplankton have been reported to show microcystin avoidance behaviour. It would be extremely interesting to learn how blister beetles and dinoflagellates avoid poisoning themselves. Is their PP1 and PP2A toxin-insensitive or have their cells evolved mechanisms to exclude toxins from the cytoplasm and nucleus?

3.   THE PHARMACOLOGICAL EFFECTS OF INHIBITION OF
      PP1 AND/OR PP2A

If okadaic acid or microcystin are applied to permeable animal or plant cells at low concentrations (~10-100 nM), a small proportion of intracellular PP1 and/or PP2A molecules are inhibited,[13,14] target proteins become more phosphorylated than normal[13,15] and the balance of intracellular control and signal responses is altered. Table 1 lists a few of the short-term consequences of treating cells, tissues and cell-free systems with low concentrations of the toxins. In many of these cases, such as activation of glycogen phosphorylase in liver cells,[16] we already had a detailed understanding of the mechanisms by which inhibition of PP1 and/or PP2A would cause the effects.[1] In contrast, the effects of the toxins were among the first experimental evidence that PP1 and/or PP2A were likely to be involved in the acquisition of memory,[17] the responses of plants to light[14,18,19] and the control of cell division[20] and viral replication.[21] Following up these observations and identifying the precise roles of protein phosphorylation in regulating these processes is under intense investigation in laboratories throughout the world. Because these toxins are such powerful pharmacological tools in medical and plant research they are sold by a number of biochemical suppliers and rank amongst their highest selling products.

When cells that are freely permeable to toxin are exposed to higher toxin concentrations (~1µM) all of the target proteins of PP1 and PP2A in a cell become fixed in their phosphorylated state.[13,15] As a result, cells are rendered unresponsive to signals and their energy supplies and structural components, quite literally, collapse.

Table 1 A selection from the hundreds of examples of
physiological effects of protein phosphatase inhibitors

Cell type or extract	Inhibitors	Effect
Smooth muscle	okadaic acid	contracts[22]
Mouse skin	okadaic acid	promotes tumour formation[23]
NIH-3T3 cells	okadaic acid	causes reversion of transformed phenotype[24]
SV40 virus T antigen	okadaic acid	inhibits DNA replication[21]
Xenopus oocyte extract	okadaic acid	activates maturation promoting factor[20]
Hepatocytes	microcystin	activates glycogen phosphorylase[16]
Rat adipocytes	okadaic acid	inhibits fatty acid synthesis and stimulates lipolysis[13]
Rat brain (hippocampus)	okadaic acid, calyculin A, microcystin*	inhibits induction of long term depression of potentiation
Paramecium	okadaic acid	prolongs backward swimming[12]
Spinach leaf	okadaic acid, microcystin	inhibits sucrose biosynthesis[19]
Wheat leaf	okadaic acid, calyculin A	inhibits light-induced greening[18]
Soybean cotyledons	all those in Fig 2	stimulates iosflavanoid production[25]

* cells were impermeable and were artificially loaded with microcystin

4.   PHYSICO-CHEMICAL NATURE OF THE INTERACTION BETWEEN MICROCYSTIN AND PP1 AND PP2A

Purified catalytic subunits of PP1 and PP2A from rabbit
skeletal muscle are potently inhibited by microcystin-LR, 50%
inhibition of either enzyme occurring at ~0.1 nM microcystin-
LR when assays are performed at phosphatase concentrations of
0.2 mU/ml using ^{32}P-labelled glycogen phosphorylase as
substrate (Figure 3).[3]   The low IC_{50} value (concentration
giving 50% inhibition) for microcystin-LR is similar to the
concentrations of PP1 and PP2A in the assays, demonstrating
that the toxin-phosphatase interactions are extremely strong.
        Consistent with essentially stoichiometric binding, the
IC_{50} values for inhibition of PP1 and PP2A increase with
increasing concentrations of phosphatase in the assay(shown in
Figure 4).[3 and similar unpublished data]

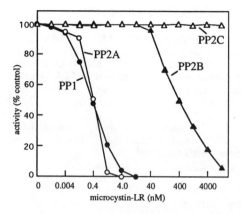

<u>**Figure 3**</u> Inhibition of rabbit skeletal muscle PP by microcystin-LR

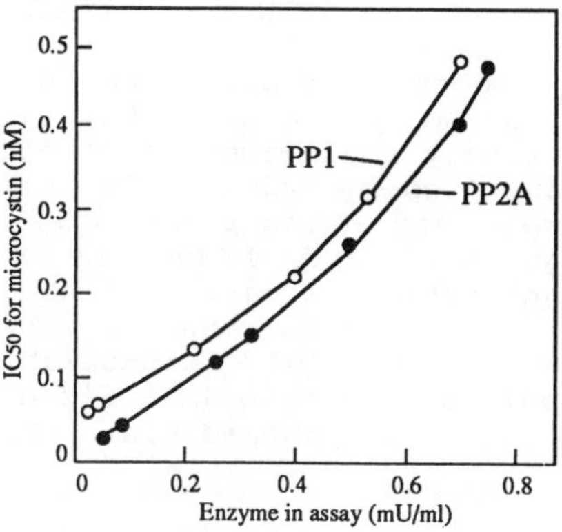

<u>**Figure 4**</u> IC_{50} values for inhibition of PP by microcystin

The nature of the strong, stoichiometric binding of microcystins to PP1 and PP2A was explored using ^{125}I-labelled microcystin-YR. The interaction between ^{125}I-labelled microcystin-YR and PP1 or PP2A was found to be stable to boiling in sodium dodecyl sulphate and precipitation in trichloroacetic acid (unpublished observations). These results are consistent with the data of Robinson *et al.*[26] who observed that ^{3}H-microcystin-LR radiolabelled a protein with an apparent molecular mass of about 40 kDa, which is close to the molecular mass of the catalytic subunit of PP1 and/or PP2A (both ~37 kDa) plus microcystin (an additional ~1 kDa). The stability of toxin-enzyme binding suggested that the interaction between the microcystins and phosphatases was covalent. By chemical cleavage, proteolytic digestion and amino acid sequencing of ^{125}I-microcystin-YR labelled protein phosphatase catalytic subunits, the microcystin binding site has been pinpointed to a 15 amino acid residue peptide located towards the C-terminus of these enzymes (R. MacK. and David Campbell, unpublished results). By aligning the amino acid sequences of all of the protein phosphatases from mammals, *Drosophila* (fruit flies), plants and yeast[refs cited in 11] that are inhibited by microcystin, it is clear that there are only one or two possible amino acids in this region that would be able to form a covalent bond with microcystin.

It has been reported that even when there is strong circumstantial evidence that an animal has died from acute microcystin poisoning, the cause of death can sometimes not be established unequivocally, as not enough microcystin is detected in the liver to account for the death. This apparent contradiction can be explained by the finding that microcystin binds covalently to protein phosphatases and, therefore, the toxin would not be present in its free form in a tissue extract, unless all of the available protein phosphatase was saturated.

5.   PROTEIN PHOSPHATASE ASSAY FOR MICROCYSTINS

An important consequence of the discovery of the molecular mechanism of microcystin action is the realisation that extremely small amounts of microcystin can now be detected and quantified by testing for inhibition of PP1 and/or PP2A in a simple and quick bioassay. The inhibition curves shown in Figures 3 and 4 may be used as calibration curves for quantitation of microcystins in this assay.

Protein phosphatase activity is determined by measuring the release of acid-soluble ^{32}P-radioactivity from ^{32}P-labelled substrate in a fixed time period:[27,28]

$$[^{32}P]\text{phosphoprotein} \xrightarrow{\text{PP1 or PP2A}} \text{protein} + \\ + nH_2O \qquad\qquad\qquad n[^{32}P]\text{phosphate}$$

The reaction is stopped by adding trichloroacetic acid (TCA) to inactivate the protein phosphatase and precipitate the unused ^{32}P-labelled protein. The acid-soluble fraction is extracted into acid molybdate which specifically extracts the inorganic phosphate and ^{32}P-phosphate is determined in a scintillation counter.

This protein phosphatase assay for detection of microcystin has the following features:

simple and quick   One person can easily perform 100 assays in a day with little training. Semi-automated assays are possible.

extremely sensitive   Less than 1 picogram of microcystin in a 100 microlitre sample can be detected.
At least a million-fold more sensitive than the rodent bioassay.
Several thousand-fold more sensitive than HPLC methods.

quantitative   Calibration curves are established.

bioassay   Inhibition of PP1/PP2A is the direct cause of the toxic effects of microcystin. Significant levels of <u>any</u> PP inhibitor in drinking water can be investigated by this method.

versatile   Drinking water, algal scum extracts and HPLC fractions can all be tested. The assay can also be used to detect okadaic acid in extracts of shellfish. Okadaic acid and microcystin can easily be distinguished from each other in this assay because these toxins have different relative potencies towards PP1 and PP2A.

For the most foolproof assay method we suggest the following procedures:-

1.   Two sets of assays should be performed, one with PP1 and one with PP2A as enzyme.   This allows double checking of microcystin concentration on both the PP1 and PP2A standard curves.[3]

2.   Phosphorylase is a recommended [^{32}P]phosphoprotein substrate because it is a good substrate for both PP1 and PP2A and is easy to make and standardise.[28]

3.   The acid molybdate extraction step should be used because there may be proteolytic enzymes present which could be mistaken for protein phosphatase activity if they are capable of releasing small TCA-soluble phosphopeptides from the phosphoprotein substrate.[28]

4.   Control blanks and standard solutions of microcystin should be assayed in every set of determinations.   Our standard solutions of stock microcystins and nodularins are quantified by amino acid analysis.[3] We recommend this method for the preparation of international standards because methods based on absorbance coefficient are prone to error.

Please note that the curves shown in Figures 3 and 4 have been established for the purified catalytic subunits of PP1 and PP2A from rabbit skeletal muscle and using 10 μM ^{32}P-labelled phosphorylase as substrate.[3]   Protein phosphatases from other sources may differ in their sensitivity to the toxins because the native enzymes consist of catalytic subunits complexed with regulatory subunits which modify protein phosphatase activity.[1]   Also, new calibration curves should be established if an alternative substrate is used.

In addition to the suggested phosphorylase phosphatase assay there are alternative possibilities using non-radioactive substrates (such as *para*-nitrophenylphosphate)[29] which may be suited to laboratories which are poorly equipped for scintillation counting.   These assays are ~20-fold less sensitive than the standard assay (C. MacK., unpublished), may be more prone to error and, as far as we are aware, have not been assessed using field samples.

6.   PROTEIN PHOSPHATASE INHIBITION:
IMPLICATIONS FOR HEALTH, WILDLIFE AND AGRICULTURE

Are there safe limits of exposure to microcystin or okadaic acid? How is wildlife and agriculture affected? High doses of microcystin have extremely nasty effects and are often lethal.   But, does chronic low-dose exposure pose a threat?   Okadaic acid and microcystin have both been shown to be tumour promoters in rodents.[23,30,31]   This is not surprising in view of the functions of PP1 and PP2A in cell growth control.   However, as this article has described, PP1 and PP2A control many processes in cells, so we should not forget that microcystin might have the potential to contribute to or exacerbate other health problems.[32]

So what are the safe limits for exposure to microcystin ? Clearly, the answer to this question depends on how many PP1 and PP2A molecules must be inactivated by microcystin over a period of time to cause changes in cellular function. We know that in experiments performed on a time-scale of minutes or hours, microcystin has obvious effects on the functions of plant and animal cells at concentrations as low as 3-10 nM (equivalent to ~3-10 μg for an adult female liver). In cells which take up microcystin freely, the maximum effects are seen at concentrations of around 1 μM, the point at which all of the cellular PP1 and PP2A is saturated with toxin. This means that approximately 1 mg (equivalent to drinking to two litres of water per day at 32 μg/ litre microcystin over two weeks) would bind all of the PP1 and PP2A in an adult female human liver, <u>provided that the PP-microcystin complexes were stable</u>. However, we actually know very little about the rates of synthesis and degradation of individual PP molecules, nor do we know whether microcystin-PP complexes accumulate in tissues or are degraded, nor whether there is a threshold dose above which cellular damage is irreparable. These are questions which are amenable to experimental investigation.

So far, most of the available data about uptake and turnover of microcystins has been obtained from experiments carried out with rodents. In this regard, it should be noted that PP1 and PP2A from rats and humans are 100% identical their amino acid sequences.[33,34] This means that information gained from animal and cell model experiments will be directly relevant to humans.

Plants come in contact with microcystins in the natural environment and on agricultural land. For example, farmers have been observed to water their crops with microcystic river water (personal observations). Our studies (for example, see Table 1) suggest that in addition to the care that must be taken to prevent the introduction of these toxins into the human food chain, it is likely that microcystin has herbicidal activity on crops and further research into this possibility seems warranted.

ACKNOWLEDGEMENTS

This work was supported by the Medical Research Council under the auspices of the MRC Protein Phosphorylation Unit. We thank our colleagues who contribute to the Unit's communal stocks of enzymes and substrates; Barry Caudwell (MRC Unit) for mass spectrophotometric analysis and, along with David Campbell (MRC Unit), Tim Gallagher (University of Bristol), Rob Field (University of Dundee) and David Gani (University of St Andrews) for invaluable advice and discussions about the chemical aspects of the work. Microcystin-YR was a generous gift from both Pieter Thiel (Research Institute for Nutritional Diseases, Tygerberg, South Africa) and Wayne W. Carmichael (Wright State University, Dayton, USA). C.MacK. holds a Fellowship from the Royal Society of Edinburgh and a project grant from the Agricultural and Food Research Council and Science and Engineering Research Council.

REFERENCES

1. Cohen, P. Annu. Rev. Biochem. 1989, 58, 453.
2. MacKintosh, C., et al. FEBS Lett. 1990, 264, 187.
3. Matsushima, R., et al. Biochem. Biophys. Res Commun. 1990, 171, 869.
4. Bialojan, C. and Takai, A. Biochem. J. 1988, 256, 283.
5. Cohen, P.et al., TIBS 1990, 15, 98.
6. Ishihara, H., et al. Biochem. Biophys. Res. Commun. 1989, 159, 871.
7. MacKintosh, C. and Klumpp, S. FEBS Lett. 1990, 277, 137.
8. Li, Y.-M. and Casida, J.E. Proc. Natl. Acad. Sci. USA 1992, 89, 11867.
9. Li., Y.-M., MacKintosh, C. and Casida, J.E. Biochem. Pharmacol. 1993, 46, 1435.
10. Eriksson, J.E., et al. Biochim. Biophys. Acta 1990, 1025, 60.
11. Cohen, P.T.W. et al. FEBS Lett. 1990, 268, 355.
12. Klumpp.S., Cohen, P. and Schultz, J.E. EMBO J. 1990, 9, 685.
13. Haystead, T.A.J. et al. Nature 1989, 337, 78.
14. MacKintosh, C. Biochim. Biophys. Acta 1992, 1137, 121.
15. Eriksson, J.E., et al. Biochem. Biophys. Res. Commun. 1990, 173, 1347.
16. Runnegar, M.T.C., et al. FEBS Lett. 1990, 264, 187.
17. Mulkey, R.M., Herron, C.E. and Malenka, R.C. Science 1993, 261, 1051.
18. Sheen, J. EMBO J. 1993, 12, 3497.
19. Siegl, G., MacKintosh, C. and Stitt, M. FEBS Lett. 1990, 27, 198.
20. Felix, M.A., Cohen, P. and Karsenti, E. EMBO J. 1990, 9, 675.
21. Lawson, R.E., Cohen, P. and Lane, D.P. J. Virol. 1990, 64, 2380.
22. Bialojan, C., et al.J Physiol., 1988, 398, 81.
23. Suganuma, M., et al. Proc. Natl. Acad. Sci. USA 1988, 85, 1768.
24. Sakai, R., et al. Proc. Natl. Acad. Sci. USA 1989, 86, 9946.
25. MacKintosh, C., Lyon, G.D. and MacKintosh, R.W. Plant J. (in press).
26. Robinson, M.A., et al. J. Pharmacol. Exp. Ther. 1991, 256, 176.
27. Cohen, P., et al. Methods. Enzymol. 1988, 159, 390.
28. MacKintosh, C. 'Protein Phosphorylation: A Practical Approach' (Hardie, D.G. ed.) IRL, Oxford, 1993, Chapter 9, p 197.
29. Takai, A. and Mieskes, G. Biochem. J. 1991, 275, 233.
30. Nishiwaki-Matsushima, R., et al. J. Cancer Res. Clin. Oncol., 1992, 118, 420.
31. Falconer, I.R. Environ. Toxicol. Water Qual. 1991, 6, 177.
32. Elder, G.H., Hunter, P.R. and Codd, G.A. The Lancet 1993, 341, 1519.
33. Barker, H.M., et al. Biochim. Biophys. Acta 1993, 1178, 228
34. Sasaki, K., et al. Jpn. J. Cancer Res. 1990, 81, 1272.

Detection of Hepatotoxins by Protein Phosphatase Inhibition Assay: Advantages, Pitfalls, and Anomalies

A. T. R. Sim and L.-M. Mudge

THE NEUROSCIENCE GROUP, FACULTY OF MEDICINE, THE UNIVERSITY OF
NEWCASTLE, CALLAGHAN, NEW SOUTH WALES, AUSTRALIA

1 INTRODUCTION

The knowledge that the hepatotoxic cyanobacterial toxins of the microcystin and nodularin classes produce their toxic effects through the specific inhibition of the cell regulatory enzymes, protein phosphatases 1 and 2A (1) has generated interest in developing assays, based on the inhibition of these enzymes, for toxin analysis. In principle, this functional-based approach should have the advantages of simplicity, sensitivity and speed when compared to other approaches such as mouse bioassay or HPLC, without the necessity for absolute toxin identification. Indeed the specificity of action of these toxins is such that it is reasonable to expect that any compound which inhibits protein phosphatases would be toxic. Such a hypothesis is supported by the fact that compounds of the okadaic acid class, which are structurally quite different from the microcystins, also specifically inhibit protein phosphatases 1 and 2A (2) and are responsible for diarrhetic shell-fish poisoning, with symptoms identical to those produced by microcystins (3). Furthermore, there are now >50 known structural variants of microcystin, including a number of hydrophobic forms (4) which have the potential to cross the cell membrane of many tissues in addition to the liver, such that it is unreasonable to expect analytical methods based on toxin identification to provide the sole basis for the detection of toxins of this class. Assays based on the inhibition of protein phosphatases therefore hold the most promise for routine analysis of such toxins and a number of laboratories worldwide have established protein phosphatase inhibition assays for this purpose. However, such an approach has not yet been widely accepted by the industry and standard procedures have not been endorsed. This may in part be due to the fact that there appears to be some variability in results obtained using this method and other approaches and between laboratories using this assay. However, such conclusions are based on limited evaluation studies (5, and Campagna, personal communication) and reflect more on the standards, methods of extraction and sample preparation used, rather than the assay itself. Furthermore, protein phosphatase inhibition assays may be carried out in a number of different ways and variability does not necessarily reflect limitations of the principle. Nevertheless, such variation requires investigation before standard procedures can be adopted. We report here findings from our own analysis of toxic and sub-toxic water samples which may in part explain some of the variability currently being encountered.

2 MATERIALS AND METHODS

GS_{1-12} peptide was purchased from Auspep and protein kinase C was a generous gift from Dr Philip Robinson. The catalytic subunit of PP2A was partially purified from rat liver by DEAE-sepharose and polylysine chromatography. Algal samples were collected as part of a routine toxin analysis service. Samples were prepared and analysed for protein phosphatase inhibition as described (6). For peptide assays, GS_{1-12} peptide (80µg/ml) was phosphorylated (to approximately 0.8 moles phosphate/ mole peptide) using 1mM ATP (containing 3µCi ^{32}P-ATP), 20mM HEPES, 1.1mM $CaCl_2$, 1mM DTT, 10mM $MgCl_2$, 0.2mg/ml PS for 6 hours at 37°C. The reaction was terminated by the addition of 30% acetic acid and the peptide was separated from ATP by chromatography on Dowex AG1X8. The peptide was dried under vacuum, neutralised, redried and resuspended in phosphatase assay buffer. Dephosphorylation of GS_{1-12} peptide was carried out as for phosphorylase phosphatase assays using purified PP2A, except that reactions were terminated by the addition of 30% acetic acid. The peptide was separated from free ^{32}Pi using Dowex AG1X8 and phosphatase activity monitored by the decrease in peptide bound radioactivity. Identification of free phosphate was by acid-molybdate extraction essentially as described by MacKintosh (7). HPLC analysis used a C18 reversed phase column, with an acetonitrile/TFA gradient as the mobile phase and microcystins were detected by the UV absorbance at 238nm.

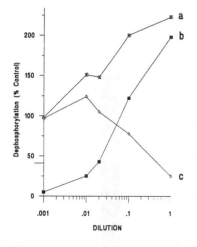

Figure 1. Effect of endogenous algal enzyme activity on toxin analysis.

Each point represents the mean of triplicate assays in which the error was always <5%. Results are expressed as the % dephosphorylation relative to a control (100%). Trace (a) is the standard measurement in the presence of exogenous phosphatase and algal extract. Trace (b) is the measurement of dephosphorylation obtained with algal extract alone. Trace (c) represents the real inhibition due to endogenous toxin and was obtained by subtraction of trace (b) from trace (a). (Redrawn from ref 6.)

3. RESULTS AND DISCUSSION

Figure 1 shows results of phosphorylase phosphatase inhibition assay obtained from 1 sample. While the presence of toxin should have been indicated by a decrease in protein phosphatase activity, a significant increase in activity was observed. Subsequently, it was shown that the cyanobacterial sample itself produced significant phosphorylase dephosphorylation, and this was masking the presence of toxin. Thus, when activity due to the "endogenous algal phosphatase" was taken into account (by subtraction of the 2 curves), there were indeed low levels of toxin observed (6). While the unavailability of further samples has prevented investigating further the nature of this particular cyanobacterial activity, we have now observed similar activity in approximately 30 different samples. While electrophoretic analysis of the phosphorylase used to assay the original sample indicated that the substrate protein was intact after exposure to the sample, we now report that proteolysis is a potential contributing factor to similar findings in at least some samples. Figure 2 shows the results of toxin analysis of another sample in which protein phosphatase activity was measured by either standard TCA precipitation of substrate and scintillation counting of the supernatant or by acid-molybdate extraction of free radiolabelled phosphate . The latter technique is specific for measuring free-phosphate while the former method will also measure phosphate bound to small peptides which may arise from proteolysis of the substrate protein. It is clear that the standard analysis (by TCA precipitation) indicated the presence of considerable "phosphatase" activity in the algal sample alone. The combination of algal phosphatase and exogenous PP2A (each of which produced approximately 30% substrate dephosphorylation) was not additive and produced activity measurements equivalent to that of the algal phosphatase alone, consistent with the presence of toxins which inhibit PP2A. However, extraction of free-phosphate in parallel assays, showed that <10% of the "phosphatase" activity measured with algal sample alone was due to the release of free-phosphate. Thus the major effect was not due to endogenous algal protein phosphatase activity. (By this method the activity of exogenous PP2A was also completely inhibited by the algal sample, again consistent with the presence of toxin).

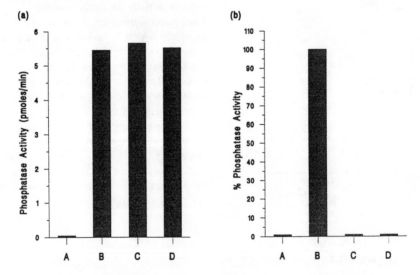

Figure 2   Results from sample analysis obtained using either (a) standard TCA precipitation or (b) acid-molybdate extraction procedures. Results are the means of triplicate determinations in which error was <5%. In each case, (A) represents the blank, (B) is the control phosphatase activity, (C) is the phosphatase activity measured in the presence of algal sample alone while (D) is the activity measured in the presence of phosphatase and algal sample.

In the majority of the samples we have now analysed, the "endogenous activity" could be diluted out to a point where it did not have an effect and toxin could be detected. Furthermore, provided appropriate controls are included, and/or the acid molybdate extraction procedure is used, we find that such endogenous algal activity does not present substantial problems to the phosphorylase phosphatase-based assay. However, there are a number of different ways in which protein phosphatase inhibition may be measured. Figure 3 shows the results of analysis of the same sample using a peptide substrate in place of glycogen phosphorylase. This approach has the advantage that phosphorylated peptides may be chemically synthesised commercially and therefore substrate preparation in the laboratory is not necessary. The results show that the peptide was effectively dephosphorylated by 5mU purified PP2A, but was relatively unaffected by concentrations of algal sample which have the equivalent of 5mU phosphorylase phosphatase activity. Furthermore, the presence of toxin is clearly indicated by the lack of dephosphorylation by PP2A in the presence of algal sample. Thus the use of the GS_{1-12} peptide substrate provides a relatively simple alternative to glycogen phosphorylase in these assays and one which appears insensitive to proteolysis. Irrespective of the method used, by taking into account the endogenous cyanobacterial phosphatase or proteolytic activity, we now generally find good agreement between HPLC and phosphorylase phosphatase inhibition assays where microcystin-LR is the toxin of interest.

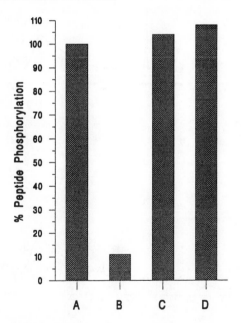

Figure 3. Effect of cyanobacterial toxins on dephosphorylation of a peptide substrate by PP2A. Results are the means of triplicate determinations where the error was <5%. (A) is the maximum peptide phosphorylation level while (B) is the phosphorylation remaining after treatment with purified PP2A. (C) is the peptide phosphorylation obtained when exposed to algal sample alone while (D) represents the level remaining after treatment with PP2A and sample.

However, in figure 4 it is shown that HPLC does not necessarily indicate the presence of toxin. In this sample which was found to have the equivalent of 32mg microcystin-LR per litre water (estimated by phosphorylase phosphatase assay using both TCA precipitation and acid-molybdate extraction methods), no substantial peak at 238nm , except that eluting in the void volume, was observed. This breakthough peak was found to be toxic in terms of protein phosphatase inhibition assay, but the amount of inhibitory activity present in this peak was insufficient to account for the majority of activity present in the sample. The nature of the active agent is currently under investigation, but remains unknown. However, since the toxic effects of microcystin and similar toxins are all likely to be due to inhibition of protein phosphatases, we assume that this compound is also toxic. Such findings therefore highlight the power of protein phosphatase inhibition assays over HPLC to detect any toxin which inhibits protein phosphatases.

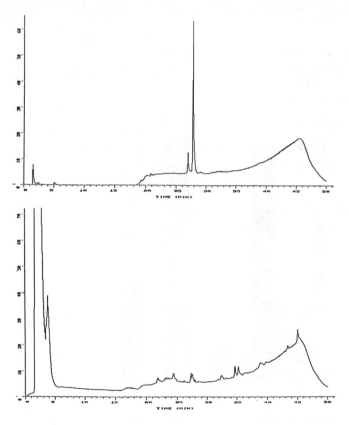

Figure 4. HPLC analysis of an algal sample containing high levels of protein phosphatase inhibiting activity. Each trace is representative of the absorbance profile at 238nm obtained using either purified microcystin-LR (upper trace) or algal sample (lower trace). Each sample was analysed 5 times and failed to show signficant levels of microcystins.

4. CONCLUSIONS

In the light of these findings we conclude that, provided appropriate controls are undertaken, protein phosphatase inhibition assay is the most effective and reliable means of detecting all hepatotoxic cyanobacterial toxins. Assays based on peptide substrates potentially provide considerable advantages over the more widely accepted protein-based assays which use glycogen phosphorylase as substrate. Modification of such assays to use non-radioactive material holds substantial promise for the development of a field kit to monitor these toxins.

5. ACKNOWLEDGEMENTS

We are grateful to the Hunter Water Corporation and the Urban Water Research Association of Australia for support.

6. REFERENCES

1. MacKintosh, C., Beattie, K.A., Klumpp, S., Cohen, P and Codd, G.A. FEBS Lett., 1990, 264, 187.

2. Hardie, D.G., Haystead, T.A.J & Sim, A.T.R. Methods. Enzymol, 1991, 201, 469.

3. Cohen, P., Holmes, C.F.B & Tsukitani, Y. TIBS, 1990, 15, 98.

4. Holmes, C.F.B. This volume, p. 85-89.

5. Falconer, I.R. In "Final Report - Appendices, Blue-green Algae Task Force " .

6. Sim, A.T.R & Mudge, L-M. Toxicon., 1993, 31, 1179.

7. MacKintosh, C. "Protein Phosphorylation, A Practical Approach ". IRL Press, Oxford, 1993, Chapter 9, p197.

Alternatives to the Mouse Bioassay for Cyanobacterial Toxicity Assessment

U. K. Swoboda,[1] C. S. Dow,[1] J. Chaivimol,[1] N. Smith,[1] and B. P. Pound[2]

[1]DEPARTMENT OF BIOLOGICAL SCIENCES, UNIVERSITY OF WARWICK, COVENTRY CV4 7AL, UK

[2]SEVERN TRENT WATER, BIRMINGHAM, UK

An optimal assay for toxicity assessment must be fast, reproducible, inexpensive and if possible avoid the use of animals. Mouse bioassays have been used as the primary means of detecting cyanotoxicity but they involve the use of live animals and pose ethical problems. We have examined several alternatives to determine cyanobacterial toxicity. These include:

a) human liver cell line, Hep G2
b) *Drosophila melanogaster* bioassay
c) a bioluminescence assay
d) polyclonal antibodies raised against microcystin LR
e) phosphatase assay.

a) Several laboratories have used freshly isolated hepatocytes in suspension as an *in vitro* system for toxicity screening of cyanotoxins and cyanobacterial blooms[1-2]. As an alternative, a hepatoma derived cell line, HepG2, was examined. Under *in vivo* conditions blebbing of the hepatocyte cell membrane occurs as a result of direct interaction of the toxin with the cytoskeleton or as a consequence of an increase in cytosolic free calcium concentration and a thiol depletion. Electron microscopic studies indicated that *in vitro* exposure of HepG2 to up to 200 μM (1-200 μg/ml) of hepatotoxins had no obvious effect on the cells. There was no significant increase in the disruption of either the cytoskeletal filaments or cell membrane over the controls following exposure of cells to the toxins for as long as 72 hours. In contrast, intraperitioneal injection of just 3 μg of microcystin LR was sufficient to kill a 20 g mouse, with an average liver weight of 1.4 g, within 20 minutes. To substantiate this observation, the uptake and release of ^{51}Cr was monitored from HepG2 cells exposed to peptide toxins, since this technique has been reported to be a good indicator of membrane damage[3]. At the same time, the release of lactate dehydrogenase (LDH) by these cells was also measured since this is also an established marker of cytotoxicity in terms of membrane damage. The data on % ^{51}Cr release indicated that there was no cytotoxic effects by microcystin LR, RR or nodularin at concentrations up to 200 μM (200 μg/ml) and for incubation periods greater than 24 hours, since the levels of ^{51}Cr released by the toxin treated cells were not significantly higher than those released by the controls. Similarly, the level of LDH released by toxin treated cells remained unchanged even with increased concentrations of the toxins and exposure time, indicating a good correlation between the LDH and ^{51}Cr release. The failure to observe cell damage and blebbing *in vitro* using HepG2 may be due to: i) the failure to take up the toxins,eg Runnegar *et al* (1991)[4] have indicated that uptake of microcystins may be, at least in part, by a carrier mediated transport system rather than by simple diffusion ii) the toxic effect may be dependent on the bioactivation of the toxin *in vivo* so that only a metabolite of the peptide toxin is the active compound and iii) *in vitro* conditions may not support the cascade of other events that precede the toxic effects.

b) A more useful indicator of cytotoxicity was achieved using the fruit fly *Drosophila melanogaster* as an assay system. Protein phosphatases are a heterogeneous group of enzymes of almost ubiquitous occurrence that are involved in reversing the action of protein kinases. Okadaic acid, a non-phorbol ester tumor promoter, produced by *Porocentrum lima*, is a potent inhibitor of type 1(PP1) and type 2A (PP2A) protein phosphatases in organisms as diverse as mammals, fruit flies, starfish, yeast and higher plants[5] . Since microcystin LR is also a potent inhibitor of these two enzymes, it seemed a possibility that the sensitivity of these enzymes in the fruit fly to microcystin LR may be virtually identical. Known concentrations of toxins or 60µl of disrupted cell biomass, as prepared for mouse bioassay, were mixed with sucrose to a final concentration of 1% and spotted onto Whatman no.1 filter discs placed at the bottom of a plastic vial. 15-30 flies which had been starved overnight were transferred to each vial. The number of flies per experiment was kept constant +/- 2 flies. Control flies had 1% sucrose in phosphate buffered saline. Death of flies was monitored regularly. Figure 1 shows that both nodularin and microcystin LR were toxic to the flies with the death rate being linked to toxin concentration and exposure time. There was a perfect correlation between the *Drosophila melanogaster* and mouse bioassays when cyanobacterial blooms produced hepatotoxins (Table 1). On the other hand, a neurotoxic *Aphanizomenon* bloom from Swithland reservoir as assessed by mouse bioassay had no toxic effect on the fruit flies as expected.

The *Drosophila melanogaster* assay is cheap, requires little sample preparation, is relatively quick, easy to do and avoids the use of mice. However, it is only useful as an indicator of toxicity and any detailed information such as dose response, would require it to be used in conjunction with mouse bioassays.

c) Antibodies were raised against purified microcystin LR by intradermal injections of rabbits[6]. Three rabbits were immunised with the conjugate and the antibody titre of the serum was determined using the competitive ELISA technique. The rabbits started to elicit antibodies after week 6 of immunisation and the titre increased after each booster injection.

An enhanced chemiluminescence (ECL) system was used to detect microcystin LR seeded into water and concentrated as for HPLC. ECL is used to detect immobilised specific antigens conjugated indirectly to horseradish peroxidase (HRP) labelled antibody. HRP catalyses the oxidation of luminol (substrate) in the presence of hydrogen peroxide. Immediately following oxidation the luminol in an excited state decays to the ground state via a light emitting pathway. The whole procedure is completed within 4-6 hours. Levels as low as 1 µg of microcystin can be detected easily using unpurified serum antibodies.

Figure 1 *Drosophila melanogaster* bioassay against nodularin and microcystin LR

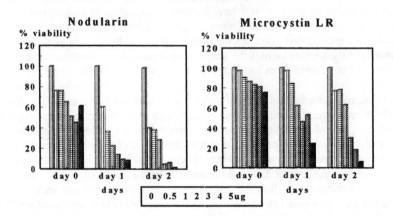

Table 1   Toxicity assessment by mouse and *Drosophila* bioassay.

Reservoir	Date	Species	Toxicity		Symptoms
			Drosophila	**Mouse**	
L Shustoke	10/6	*Oscillatoria*	toxic	toxic	H
U Shustoke	2/9	*Oscillatoria*	toxic	toxic	H
Cropston	19/8	*Microcystis*	toxic	toxic	H
Swithland	26/6	*Aphanizomenon*	non-toxic	non-toxic	
Swithland	14/10	*Aphanizomenon*	non-toxic	toxic	N
		Microcystis PCC7806	toxic	toxic	H

| H : hepatotoxin   N : neurotoxin |

Since this procedure is potentially far more sensitive and has the ability to detect much lower levels of antigens, it is being refined and its suitability to detect (i) microcystin variants and other hepatotoxins and (ii) toxic environmental cyanobacterial samples and contaminated raw water are being assessed. This system has the advantage in that it is non-radioactive, quick, able to detect small amounts of antibodies and produces stable hard copy data.

d)   The use of bacterial bioluminescence based Microtox assays using *Photobacterium phosphoreum* was also evaluated for cyanobacterial toxicity assessment. Three purified toxins, 1 laboratory isolate and 9 natural blooms of cyanobacteria were tested and the toxicity data compared with mouse bioassay results. Samples were prepared for the assay as described by Lawton, L.A. *et al*[1] and the assay carried out according to the manufacturer's operating instructions (Microbics Corporation, USA).

The effective concentration of sample causing a 50% decrease in light emission, EC50, was then calculated. The EC50 of microcystin LR, microcystin RR, nodularin and anatoxin-a were 0.102, 0.097, 0.147 and 0.164 mg/ml while their lethal dose were 3,30,3 and 5 μg/20g mouse respectively. If toxicity is inferred at EC50 < 0.5 mg/ml (value set by Lawton *et al*), then only 1 out of the 7 hepatotoxin producing strains (as determined by by mouse bioassay and HPLC) would be classified as toxic while 2 of the 4 non-toxic strains would be deduced to be toxic (Table 2). If however toxicity was inferred at a much higher value of EC50 ie, <1 mg/ml , then 5 out of the 7 hepatotoxin producing strains could be deduced as being toxic but 3 out 4 non-toxic strains would be deduced as being toxic. Also the same strain of *Oscillatoria* sp. from Lower Shustoke collected at different times, shown to be toxic by mouse assay, had lethal doses of 83.5 and 400 mg dry wt cell lysate/kg body weight respectively. Surprisingly, the former more potent sample was deduced as being non-toxic while the latter as toxic. Thus, the lack of correlation between different samples and the risk of false negatives/positives indicates that the Microtox is not a reliable tool for detecting cyanobacterial toxicity. Other disadvantages include: (i) the sensitivity to pure toxins is relatively low (~100μg compared to 10 ng by HPLC) and there is no specificity (ii) the sensitivity of the Microtox to other chemicals eg chlorine, does not make it a suitable tool for assessing treated water.

Table 2   Comparison of toxicity assays by mouse bioassay and bioluminescence.

Sample	Source	Mouse bioassay		Bioluminescence		
		LD (1)	Tox (2)	EC50(3)	Tox(4)	Tox(5)
A cylindrica	laboratory	>1500	nt	>5	nt	nt
M aeroginosa	PCC7806	55	toxic	0.495	toxic	toxic
M aeroginosa	Cropston	<100	toxic	0.98	nt	toxic
Oscillatoria	L Shustoke	83.5	toxic	1.39	nt	nt
Oscillatoria	L Shustoke	400	toxic	0.93	nt	toxic
Oscillatoria	Linacre	50	toxic	0.82	nt	toxic
Oscillatoria	Earlswood	<125	toxic	1.08	nt	toxic
Oscillatoria	Thornton	195	toxic	1.325	nt	nt
O. erythrea	Australia	>1500	nt	0.215	toxic	toxic
Aphanizomenon	Swithland	>1000	nt	0.525	nt	toxic
Aphanizomenon	Swithland	>1000	nt	0.485	toxic	toxic

1 : mg dry wt/kg body weight   2 : death within 24 hours of intraperitoneal injection
3 : mg dry wt/ml   4 : inferred from EC50 <0.5mg/ml   5 : inferred from EC50 <1 mg/ml
nt : non-toxic

e)   One of the most promising assays to determine hepatotoxicity of cyanobacterial blooms is the phosphatase assay. The assay involves the conversion of ^{32}P labelled   serine phosphorylase a to phosphorylase b using mouse liver homogenate as the source of phosphatases. During this reaction, ^{32}P released is measured. Microcystin LR and okadaic acid are potent inhibitors of protein phosphatases 1 and 2A. The potency of toxins on the enzyme is determined by its IC_{50} (concentration of toxin required to cause 50% inhibition of the phosphorylase phosphatase activity). The IC_{50} of several purified toxins and natural cyanobacterial blooms were tested for toxicity.

Microcystin LR, microcystin RR, nodularin and okadaic acid strongly inhibited phosphorylase activity. However, the cyanobacterial hepatotoxins were 10 times more toxic than okadaic acid. Microcystin LR is 10 times more toxic than microcystin RR when injected intraperitioneally into mice but they both have approximately the same IC_{50}. The difference may be due to either permeability effects and /or differential activation of the microcystins *in vivo*. Cell extracts of all environmental samples of *Oscillatoria* and *Microcystis,* shown to be toxic by mouse bioassay, were inhibitors of serine phosphatase while the non-toxic strain of *Oscillatoria* from Lower Shustoke had no effect on the enzyme. 200 ml concentrates of cell free raw reservoir water collected at the same time had no inhibitory effect on the phosphatase. Thus so far, this assay shows a good correlation to data obtained by mouse bioassay and HPLC. Details of the experimental procedure and further data are presented in an accompanying paper by Chaivimol *et al.* Further work is being carried out to assess the analytical performance with very low levels of hepatotoxin in freshwater reservoirs with high levels of cyanobacteria.

REFERENCES

1.   M.T. Runnegar, I.R.Falconer and J. Silver, Arch. Pharmac., 1981, 317, p268.
2.   M.T. Runnegar, J. Andrews, R.G. Gerdes and I.R. Falconer, Toxicon, 1987, 25, p1235.
3.   R. Zawydiwski and G.R.Duncan, In Vitro, 1978, 14, p707.
4.   M.T.Runnegar, R.G. Gerdes and I.R. Falconer, Toxicon, 1991, 29, p43.
5.   P. Cohen and P.T.W. Cohen, J.Biol Chem., 1989, 264, p21435.
6.   U.Chu, Xuan Huang, R.D. Wei and W.W.Carmichael, App. Environ. Microbiol., 1989, 55, p19329.
7.   L.A. Lawton, D.L. Cambell, K.A. Beattie and G.A. Codd, Lett. App. Microbiol., 1990, 11, p205.

ACKNOWLEDGEMENT

This research was sponsored by Severn Trent Water, Birmingham, UK.
Their help and involvement, especially that of Les Markham and Helen Picket, is gratefully acknowledged.

Evaluation of Assay Methods for the Determination of Cyanobacterial Hepatotoxicity

L. A. Lawton,[1,2] K. A. Beattie,[1] S. P. Hawser,[1] D. L. Campbell,[1] and G. A. Codd[1]

[1]DEPARTMENT OF BIOLOGICAL SCIENCES, UNIVERSITY OF DUNDEE, DUNDEE DD1 4HN, UK

[2]PRESENT ADDRESS: SCHOOL OF APPLIED SCIENCES, THE ROBERT GORDON UNIVERSITY, ABERDEEN AB1 1HD, UK

1 INTRODUCTION

The hazards posed to both humans and animals by hepatotoxic blooms of cyanobacteria (blue-green algae) are of increasing interest in both recreational and potable water supplies, necessitating the development of rapid, reliable and sensitive assay methods. The mouse bioassay has been, to date, the most extensively used to determine the toxicity of cyanobacterial blooms; however, this is both expensive and is often opposed on moral grounds. Many lower organisms and cultured animal cells have been used to develop alternative assay methods throughout the toxicological field and it would be desirable to investigate the possible use of such methods to determine the toxicity of cyanobacterial blooms.

The toxicity of 21 cyanobacterial samples was assessed by six methods of determining toxicity and the results compared to those obtained by the mouse bioassay. The additional assays assessed were: high-performance liquid chromatography with diode array detection (DAD-HPLC); brine shrimp (*Artemia salina*); Microtox bioluminescence; V79 fibroblast cytotoxicity; *Serratia marcescens* pigment (prodigiosin) inhibition and *Daphnia pulex*.

The suitability of each method is discussed, taking into account the assay's ability to determine differing levels of microcystins (or nodularin), the skill required to perform the assay and the relative cost associated with its use.

2 MATERIALS AND METHODS

The cyanobacterial samples used included species of *Microcystis*, *Anabaena*, *Oscillatoria*, *Gomphosphaeria*, *Aphanizomenon*, *Nodularia* and *Gloeotrichia*. Three of the samples were from laboratory grown cultures and the remainder were naturally occurring bloom material collected throughout the UK, including both monocyanobacterial and mixed examples. Approximately half of the samples selected were found to be hepatotoxic when tested by mouse bioassay, the method for which is described elsewhere[1,2].

The material selected for testing had been stored at -20°C after lyophilization. It was extracted twice in 5% (v/v) acetic acid (80 ml per 1 g) and the supernatants from both extractions combined and passed through a Waters C_{18} Sep-Pak cartridge. The cartridge was then eluted with methanol (20 ml) and the eluate rotary evaporated at 40°C to dryness. This was resuspended in methanol (1 ml) and divided into aliquots which were stored dry at -20°C until tested. Prior to assessment the extracts were taken up in a small volume of methanol then the appropriate medium added to give the required dilution. Controls which included the corresponding amount of methanol were carried out for each assay. It should be noted that the concentrations of extracts throughout the experiment are expressed as mg dry weight of cells extracted not the actual weight of the extract itself.

Most of the methods used are fully described in other publications; hence only a brief overview will be given here. DAD-HPLC[3,4] analysis was used to quantify and identify microcystins (and nodularin) in the samples. This was achieved by first identifying the toxins by their characteristic spectra, then all peaks indicating a positive match were quantified by peak area with respect to a microcystin-LR standard.

The brine shrimp, *S. marcescens* and Microtox bioluminescence assays are fully described in Campbell *et al.*[5], Dierstein *et al.*[6] and Lawton *et al.*[7] respectively. The cytotoxicity assay was carried out using Chinese hamster lung cell fibroblasts adopting a method which colorimetrically determines the metabolic activity of the cells using the tetrazolium salt MTT. The *D. pulex* bioassay was performed using ten neonates in 1 ml (using 24-well tissue culture plates) volumes, in triplicate. The percentage mortality was calculated after 24 h and the LC_{50} value calculated.

3 RESULTS AND DISCUSSION

The results obtained from each of the assay methods were compared to those obtained by mouse bioassay (Figure 1) and in all assays, with the exception of the Microtox, tentative toxicity ratings were derived which correlated with high, medium, low toxicity and non-toxic samples (Table 1). These toxicity ratings were used to determine the suitability of each of the assay methods. Table 2 summarizes the percentage correlation found between the assays under investigation and the mouse bioassay. Further comparison was made between the toxicity rating ± one, and the values found by mouse bioassay, to accommodate intermediary values.

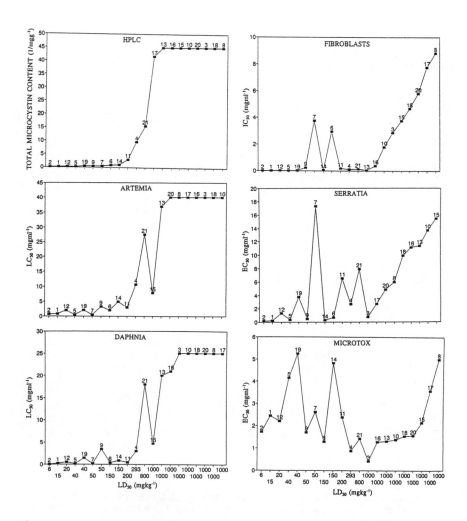

<u>Figure 1</u> Comparative plots of cyanobacterial laboratory
culture and natural bloom samples (numbered 1-21)
toxicity determined by mouse bioassay (LD_{50} mgkg^{-1})
against the toxicity values obtained from each of the
assays under investigation. To allow ease of
comparison the total microcystin content determined
by DAD-HPLC is plotted as the reciprocal
concentration (1/mgg^{-1}).

<u>Table 1</u> Tentative toxicity ratings based on the results from this study.

Toxicity rating	Degree of toxicity	Mouse bioassay (LD_{50} mgkg^{-1})	HPLC (mgg^{-1})	*Artemia* (LC_{50} mgml^{-1})	*Daphnia* (LC_{50} mgml^{-1})	V79 cells (IC or EC_{50} mgml^{-1})	*Serratia*
NT	non-toxic	>1000	<0.01	>30	>20	>1.0	>10
1	low	>500<1000	>0.01<0.1	>10<30	>5<20	>0.5<1.0	>5<10
2	medium	>100<500	>0.1<1.0	>2<10	>1<5	>0.1<0.5	>1<5
3	high	<100	>1.0	<2	<1	<0.1	<1

<u>Table 2</u> Percentage correlation of toxicity ratings compared to the mouse bioassay values.

Assay	Direct comparison (% correlation)	Comparison ± 1 (% correlation)
HPLC	91	100
Artemia	76	95
Daphnia	71	95
Fibroblasts	67	81
Serratia	48	81
Microtox	ND	ND

Further evaluation of the results was carried out by assessing the number of false positive and negative values obtained for each assay (Table 3). The presence of the occasional false positive may be expected as it is difficult to provide a bioassay which responds to only one group of toxicants. In the case of the HPLC assay the false positive occurred due to the detection of low amounts of microcystins which the mouse bioassay failed to detect. However, false negatives are undesirable as the hazard presented by a bloom would go undetected.

<u>Table 3</u> Number of false positive and negative values obtained when assay results were compared to those of the mouse bioassay.

Assay	False +ve	False -ve
HPLC	1	0
Artemia	1	0
Daphnia	1	0
Fibroblasts	2	2
Serratia	5	1
Microtox	ND	ND

The use of a simple clean-up and concentration step enabled the testing of the cyanobacterial samples at higher concentrations. The clean-up step selectively isolated the hepatotoxin-containing fraction hence reducing the chance of interference by other compounds and increasing the confidence in the assay detecting primarily cyanobacterial hepatotoxins. Recent work[3,4] has indicated that aqueous extraction fails to extract the more hydrophobic microcystin variants. Therefore the use of methanolic extraction is recommended.

HPLC analysis has become increasingly important in the analysis and quantification of cyanobacterial hepatotoxins. However with the use of single wavelength detection it is only possible to quantify microcystins if the purified standard of a particular variant present in the sample is available. The use of DAD-HPLC provides spectral information on each peak observed in a chromatogram allowing tentative identification. In the present study, DAD-HPLC analysis was found to accurately determine the hepatotoxicity of samples showing a high degree of correlation with the mouse bioassay (Figure 1, Table 2).

Both bacterial systems (Microtox and *Serratia*) were found to show little correlation with hepatotoxicity as determined by mouse bioassay. A recent study[5] has shown that the inhibitory effect observed in the Microtox system was associated with cyanobacterial components other than microcystins. The tissue culture system (fibroblast cells) also gave poor correlation with the mouse bioassay suggesting its unsuitability for the detection of microcystins. However, the invertebrate bioassays (*Artemia* and *Daphnia*) gave good correlation indicating their potential as screening methods for the presence of microcystins.

4 CONCLUSIONS

The relative merits of the six methods carried out for the detection and quantification of cyanobacterial hepatotoxins are summarized as follows:

DAD-HPLC: Expensive; degree of expertise required; quantitative and specific.

Artemia: Inexpensive; relatively easy to set up; quantitative but not specific.

Daphnia: Inexpensive; labour intensive; quantitative but not specific.

Fibroblasts: Correlation not as good as above tests; expensive; labour intensive; not specific.

Serratia: Poor correlation.

Microtox: No correlation.

In conclusion, the current investigation has shown that DAD-HPLC is the best alternative to the mouse bioassay of the six methods examined, with the additional advantage of increased sensitivity and specificity. However, the invertebrate assay systems were demonstrated to be useful, inexpensive screening methods which can be easily established in most laboratories.

ACKNOWLEDGEMENTS

SPH and DLC thank the Science and Engineering Research Council and Natural Environment Research Council, respectively, for postgraduate research studentships. We also acknowledge the kind supply of *Serratia marcescens* by Dr. David C. Old and the helpful interest of Dr. Christine Edwards.

REFERENCES

1. G.A. Codd and W.W. Carmichael, FEMS Microbiol.Lett., 1982, 13, 409.
2. D.S. Richard, K.A. Beattie and G.A. Codd, Environ.Technol.Lett., 1983, 4, 409.
3. L.A. Lawton, C. Edwards and G.A. Codd, The Analyst, 1994, 119, 1525.
4. L.A. Lawton, C. Edwards and G.A. Codd, (these Proceedings), p. 59-63.
5. D.L. Campbell, L.A. Lawton, K.A. Beattie and G.A. Codd, Environ.Toxicol. Water Qual., 1994, 9, 71.
6. R. Dierstein, I. Kaiser and J. Weckesser, System.Appl.Microbiol., 1989, 12, 244.
7. L.A. Lawton, D.L. Campbell, K.A. Beattie and G.A. Codd, Letts.Appl.Microbiol., 1990, 11, 205.

The Role of Synthetic Chemistry in the Production of Standards for Toxin Analysis

Timothy Gallagher,[1]* Paul A. Brough,[1] Terry M. Jefferies,[2] and Susan Wonnacott[3]

[1]SCHOOL OF CHEMISTRY, UNIVERSITY OF BRISTOL, BRISTOL BS8 1TS, UK

[2]SCHOOL OF PHARMACY, UNIVERSITY OF BATH, BATH BA2 7AY, UK

[3]SCHOOL OF BIOLOGICAL SCIENCES, UNIVERSITY OF BATH, BATH BA2 7AY, UK

1   WHAT ARE THE AVAILABLE SOURCES OF TOXINS?

The chemistry of algal and related toxins has been an active area of research for synthetic chemists for some years.[1] Much of this work has been motivated by a combination of both the complexity of structures involved and the biological properties that these molecules possess. As a result, it is quite clear that synthetic chemistry can make a significant contribution both now and in the future to the problems associated with the analysis of naturally derived toxins in the environment.

In terms of supplying toxins as standards for detection methods, two questions immediately arise: does Nature provide us with an adequate and accessible source of material, i.e. is isolation of the toxin from its natural source a viable proposition? If not, and this may be the case for a combination of several reasons - access to and stability of biological material being common problems - then can synthetic chemistry be used to fill this need? Frequently the answer to this second question is yes, but this statement must be qualified. At this point it is also relevant to pose a third question and ask if synthetic chemistry can provide the analyst with new, alternative tools? The answer to this is definitely yes.

Consideration must first be given to the options available for obtaining the toxins that are needed to develop, quantify and monitor analytical procedures. Nature obviously has tremendous potency. There are single organisms that incorporate the machinery needed to produce an array of different, though sometimes structurally related, toxin components.   Nature also has the biological mechanisms available for generating structurally complex molecules, such as the brevetoxins, and while synthetic chemistry has yet to prove that it can compete on equal terms with this level of sophistication, the power of modern synthetic methodology and its practitioners must not be underestimated.[2]

With structurally simpler molecules, the attractions associated with synthetic materials are both obvious and demonstrated. Single, well-defined chemical entities are available and this supply is not dependent on what can often be a fickle biological source. Good, efficient chemistry can also produce relatively large (gram) quantities of toxin, quantities that are necessary both for use in analytical method development as well as in studies needed to determine the biological effects and risks associated with these toxins. The biological information obtained from these studies clearly has an impact on the course of the development of analytical methods.

2   THE NATURE OF THE SYNTHETIC PROBLEM

Before embarking on a synthetic approach to a toxin, we must make a number of decisions. Firstly, there is the complexity of the target to consider and we must then be in a position to identify and develop a viable synthetic strategy. Such decisions can only properly be made with a comprehensive knowledge of the science and by addressing more specialised issues relating to stereochemistry. Of the classes of toxins that are of current interest, both the microcystins and the anatoxins are amenable NOW to modern synthetic methodology. Relatively little has been done in the microcystin area - we have not yet been involved in this area - but a brief overview of the chemistry being carried out in Bristol on the anatoxins will serve to illustrate some of what is now available.

3   THE CHEMISTRY OF ANATOXIN-A AND RELATED TOXINS

Appreciation of stereochemistry is an important topic but one that often leads to confusion. (+)-Anatoxin-a **1**, which is the naturally-occurring enantiomer,[3] is much more difficult to produce in the laboratory than is the racemate - (±)-anatoxin-a. Nature does not, as far as is known, produce the (-)-enantiomer,[4] but the racemate (or the unnatural (-)-enantiomer) is nevertheless ideal for use as an analytical standard since the relevant physical properties of both enantiomers are the same. The pure (+)-enantiomer of anatoxin-a is, however, required for any biological studies - anatoxin-a is a potent nicotinic acetylcholine agonist and is of neurochemical interest[5] - because under these conditions, which relates to the asymmetric constraints imposed by a receptor, the behaviour of the two enantiomers is different.

1 *Anatoxin-a*   **2** *Homoanatoxin*

What can be done in the laboratory? To date, we[6] are the only group in the UK to have published on the synthesis and further chemistry of anatoxin-a and protocols are now available that will rapidly provide gram quantities of this toxin. The problem of gaining access to pure enantiomers has also been tackled and, very recently, a clean and synthetically feasible resolution (involving HPLC methods) has been identified. This approach can certainly compete (in terms of time and costs of consumables) with previously published methods for asymmetric synthesis[7] and promises to make a significant impact on the field.

Homoanatoxin **2** - A New Algal Toxin.

There are other opportunities available in this area that synthetic chemistry is in a unique position to exploit.   Can we, based on our knowledge of structure and biological mechanisms, second guess Nature and perhaps make toxin entities in the laboratory prior to their discovery as a result of, for example, a bloom incident?   This is a somewhat contrived point because the principle incentive to develop new toxins in this area is derived from biological and neurochemical rather than analytical or environmental concerns.   However, homoanatoxin provides an example of the potential of the methods now available.   Homoanatoxin **2** is the ethyl ketone analogue of anatoxin-a **1**.[8] Homoanatoxin was first made in our laboratory (from anatoxin-a in three simple steps) and was designed as a vehicle for a [^{3}H]-radiolabel for the study of nicotinic neuronal pathways in brain.   (±)-Homoanatoxin has now been resolved into the (+)- and (-)-enantiomers by our HPLC method and displays a very similar neurochemical profile to anatoxin-a. Is homoanatoxin a naturally-occurring toxin? Homoanatoxin was isolated by Skulberg and co-workers[9] from a toxic bloom of *Oscillatoria formosa* in Norway and, based on mouse bioassay, has been shown to be as toxic as anatoxin-a.

Are There Other Toxins Waiting to be Found?

The situation encountered with homoanatoxin is somewhat unusual - it is not often that synthesis of a molecule precedes its isolation from Nature - but did raise the possibility that there might also be other toxins waiting to be made.   Two other molecules have recently been synthesised and, because of the structure of the ketone side chain, have been named as propylanatoxin **3** and isopropylanatoxin **4** respectively.   Both molecules are highly potent nicotinic ligands,[10] a property which will likely correlate with mammalian toxicity, however, neither propylanatoxin **3** nor isopropylanatoxin **4** have yet to be observed in Nature.

3 *Propylanatoxin*   **4** *Isopropylanatoxin*

The intention of this aspect of our work is not to make life more difficult for the water industry, but a serious point needs to be made.   Currently, little is being done to unravel the biosynthetic origins of this class of algal toxin.[11]   Given the appropriate carbon feedstuffs, is it feasible that propylanatoxin and isopropylanatoxin could become algal metabolites and, as such, as much of a hazard as anatoxin-a?   Knowledge of Nature's requirements to produce these products may provide an important insight into the extent of the problem that we may face.   To neglect the topic of biosynthesis not only serves to deny a valuable contribution to bioorganic chemistry but also reflects a lack of willingness to accept the possible breadth of the broader environmental problem.

4   CAN SYNTHETIC CHEMISTRY OFFER NEW
        ANALYTICAL TOOLS?

In addition to the topics described above, synthetic chemistry can provide, for example, deuterium-labelled toxins for use as internal standards in mass spectrometric methods. An immunoassay procedure would be an attractive new analytical procedure in this field and is especially appropriate as the basis of a quick and very sensitive field test. In the anatoxin area, we have demonstrated[12] the feasibility of generating the basic ligand conjugates that are required to implement such a procedure. Although further work is still required to take this to the point where a viable analytical tool is available, such a procedure would likely serve to detect the structurally related and equally toxic homoanatoxin.

5   CONCLUSIONS

In conclusion, synthetic chemistry can, especially when groups with a complementary range of analytical and biological skills come together, make a substantial and varied contribution to the problems associated with algal toxins. The development of analytical methods and the results of biological studies must be considered jointly; the toxicity of the target chemical entity must have an impact on the priorities associated with analytical goals. Standards are now available (and have been used) for both anatoxin-a and homoanatoxin[13] and broad spectrum biological evaluation, especially of new toxins, can be undertaken. While conventional analytical procedures currently dominate this field, the power of modern synthetic chemistry, in collaboration with other key disciplines, can offer access to a range of other tools that merit a closer level of examination and evaluation than has been forthcoming to date.

REFERENCES

1.   (a) W. W. Camichael, J. Appl. Bacteriology, 1992, 72, 445-459; (b) Many aspects of the chemistry of marine toxins have been reviewed recently: Chem. Rev., 1993, 93, 1671-1944.

2.   The synthesis of the complex marine polyether toxin hemibrevetoxin B has been achieved recently: K. C. Nicolaou, K. R. Reddy, G. Skokotas, F. Sato, X.-Y. Xiao and C.-K. Hwang, J. Am. Chem. Soc., 1993, 115, 3558-3575.

3.   W. W. Carmichael, D. F. Biggs and P. R. Gorham, Science, 1975, 187, 542; J. P. Devlin, O. E. Edwards, P. R. Gorham, N. R. Hunter, R. K. Pike and B. Stavric, Can. J. Chem., 1977, 55, 1376-1371. The structure of anatoxin-a was established by x-ray crystallographic analysis: C. S. Huber, Acta Crystallogr. Sect. B, 1972, 28, 2577.

4.   (-)-Anatoxin shows negligible neurotoxic activity compared to the (+)-enantiomer: K. L. Swanson, C. N. Allen, R. S. Aronstam, H. Rapoport and E. X. Albuquerque, Mol. Pharmacol., 1986, 29, 250-257; P. Kofuji, Y. Aracava, K. L. Swanson, R. S. Aronstam, H. Rapoport and E. X. Albuquerque, J. Pharmacol. Exp. Ther., 1990, 252, 517-525.

5.   (a) C. E. Spivak, B. Witkop and E. X. Albuquerque, <u>Mol Pharmacol</u>., 1980, <u>18</u>, 384-394; (b) P. Thomas, M. W. Stevens, G. Wilkie, M. Amar, G. G. Lunt, P. Whiting, T. Gallagher, E. F. R. Pereira, M. Alkondon, E. X. Albuquerue and S. Wonnacott, <u>J. Neurochem</u>., 1993, <u>60</u>, 2308-2311.

6.   N. J. S. Huby, R. G. Kinsman, D. Lathbury, P. G. Vernon and T. Gallagher, <u>J. Chem. Soc., Perkin Trans. 1, 1991</u>, 145-155.

7.   (a) A. M. P. Koskinen and H. Rapoport, <u>J. Med. Chem</u>., 1985, <u>28</u>, 1301-1309; (b) P. Stjernlöf, L. Trogen and Å. Andersson, <u>Acta Chem. Scand</u>., 1989, <u>43</u>, 917-918; (c) P. Somfai and J. Åhman, <u>Tetrahedron Lett</u>., 1992, <u>33</u>, 3791-3794.

8.   S. Wonnacott, K. L Swanson, E. X. Albuquerque, N. J. S. Huby, P. Thompson and T. Gallagher, <u>Biochem. Pharmacol</u>., 1992, <u>43</u>, 419-423.

9.   O. M. Skulberg, W. W. Carmichael, R. A. Andersen, S. Matsunaga, R. E. Moore and R. Skulberg, <u>Environ. Toxic. Chem</u>., 1992, <u>11</u>, 321-329.

10. P. Thomas, P. A. Brough, S. Wonnacott and T. Gallagher, <u>unpublished work</u>.

11. J. R. Gallon, K. N. Chit and E. G. Brown, <u>Phytochemistry</u>, 1990, <u>29</u>, 1107-1111.

12. N. J. S. Huby, P. Thompson, S. Wonnacott and T. Gallagher, <u>J. Chem. Soc., Chem. Commun., 1991</u>, 243-245.

13. A. Zotou, T. Jefferies, P.A. Brough and T. Gallagher, <u>Analyst</u>, 1993, <u>118</u>, 753-758; J.E. Haugen, O.M. Skulberg, R.A. Andersen, J. Alexander, G. Lilleheil, P. A. Brough and T. Gallagher, <u>Arch. Hydrobiol.</u>,1994, in press.

Sources of Uncertainty in Assessing the Health Risk of Cyanobacterial Blooms in Drinking Water Supplies

Steve E. Hrudey,[1] Sandra L. Kenefick,[1] Timothy W. Lambert,[1] Brian G. Kotak,[2] Ellie E. Prepas,[2] and Charles F. B. Holmes[3]

[1]ENVIRONMENTAL HEALTH PROGRAM, DEPARTMENT OF PUBLIC HEALTH SCIENCES

[2]LIMNOLOGY PROGRAM, DEPARTMENT OF ZOOLOGY, FACULTY OF SCIENCE

[3]DEPARTMENT OF BIOCHEMISTRY, FACULTY OF MEDICINE
UNIVERSITY OF ALBERTA, EDMONTON T6G 2G3, CANADA

1 INTRODUCTION

The occurrence of cyanobacterial toxins in drinking water supplies has been gaining increasing attention as the methods for the detection and identification of toxins have dramatically improved. These advances have increased our awareness of the potential health risks that these toxins may pose. Health risk assessment has also been developing over the past decade as an approach for determining the bounds to potential environmental health problems. An important aspect of health risk assessment is understanding the sources contributing to the uncertainty of the health risk estimates. We consider the general approach of health risk assessment and the specific contributions that sampling and analytical components may contribute to uncertainty in the overall health risk estimate.

2 ENVIRONMENTAL HEALTH RISK ASSESSMENT OF TOXINS.

To use risk assessment as an effective means for evaluating environmental health problems, we must first understand that it is only an approach for investigating and organizing information. Health risk assessment cannot provide all the answers nor should it become an end unto itself.

A meaningful discussion of health risk must answer: 1.What can go wrong? 2.How likely is it to happen? 3.If things do go wrong, what are the health consequences?

To answer these questions, a health risk assessment must: 1.specify or identify a scenario that poses tangible dangers or hazards, 2.estimate the likelihood of the specified scenario(s) occurring, 3.estimate the consequences to human health if the scenario does occur. For cyanobacterial toxins, a health risk assessment should seek to accurately predict the human health risks for a given toxin exposure scenario and thereby provide a rational basis for managing exposures and reducing plausible health risks.

Generally, risk management options can seek to limit health risks by: 1.eliminating the possibility of the hazardous scenarios, 2.reducing the probability of the hazardous scenario occurring, 3.mitigating the adverse consequences if the hazardous scenario does occur. These options may be illustrated for the specific case of cyanobacterial toxin health risks to drinking water supplies by, for example: 1.replacing a surface water supply prone to toxic blooms with a groundwater supply or some other source free of any potential for toxin problems, 2.reducing the likelihood of toxic blooms occurring by reducing nutrient inputs to the water supply, and/or, 3.treating the water from the source using technology that has been proven to remove the toxins present.

Each of these options must be weighed against the other risks involved (in the case of alternate water sources) and the costs incurred for the level of risk reduction that may be achieved. While risk assessment may be able to organize our understanding of the health risks of a given scenario, the practical decisions to manage risks will always require considerable judgement.

The key elements of a health risk assessment as it applies to cyanobacterial toxins are: 1.source characterization and hazard assessment, 2.human exposure assessment, 3.health consequence assessment, 4.overall health risk characterization. Our brief discussion deals only with elements 1, 2 and 4.

Source Characterization and Hazard Assessment.

For cyanobacterial toxins, we need to know their basic chemical and physical properties, their fate and behaviour properties and their toxicological properties.

The advances in characterizing toxins has allowed their study in pure form so that some information has been published on their basic chemical and physical properties. More is being learned about their fate and behaviour properties, such as water solubility, partitioning coefficients among environmental media, and stability factors for processes like biodegradation, hydrolysis and photolysis.

Basic toxicological properties should be considered in the same manner that would be applied to a new synthetic substance under current toxic substance registration legislation. Given the substantial potential for human exposure to these toxins, we should have information on their full range of toxic properties including: acute lethality by ingestion, chronic lethality and morbidity, acute and chronic dermal irritation, reproductive toxicity, mutagenic, clastogenic and teratogenic properties and finally, tumour intiation and promotion capabilities. Although, much work is being done on the toxic properties of cyanobacterial toxins, we are well short of having adequate characterization.

In our work over the past four years on cyanobacterial toxins, we have been guided by the available toxicology information. Acute human health risks might be posed by either neurotoxins or hepatotoxins, given sufficiently high exposures. However, we reason that if acute toxin poisoning were to have occurred in a drinking water supply, some attention would likely to have been raised, provided that the exposed population was large enough and the symptoms of the illness were sufficiently characteristic to generate a cause-effect hypothesis. According to our understanding of the toxic action of the neurotoxins, chronic health effects from exposures below an effective dose for acute neurotoxic effects seems unlikely. However, in the case of the hepatotoxins (primarily microcystins in our region), their mode of toxic action clearly allows for subtle chronic health effects. Given this logic and the results of our first year (1990) of screening lakes in central and northern Alberta[1], we have focused our attention on the potential health risks associated with chronic exposure to microcystins in drinking water supplies.

Exposure Assessment.

The focus of our work to date has emphasized determining the distribution, character and treatability of microcystins (primarily microcystin-LR) in bloom biomass and drinking water supplies. This work provides a basis for estimating the plausible external (ingested) dose of microcystin-LR through some drinking water supplies in Alberta. Because of the analytical focus of these proceedings, we will discuss the sources of uncertainty that arise primarily in the exposure assessment stage.

Based on over 380 bloom biomass sampled from 19 lakes over a three year period, more than 70% have contained detectable levels (> 1 μg toxin / g of dry biomass) of microcsytin-LR. Levels determined by high performance liquid chromatography (HPLC) with UV detection have ranged from barely detectable to 1550 μg/g in bloom biomass. Similarly, total microcystin activity, as measured by the phosphatase inhibition assay, have ranged from 0.15 to 4.3 μg/L in raw intake waters and from 0.09 to 0.64 μg/L in treated drinking waters.

Sample Collection and Preparation.

All cyanobacterial bloom samples were collected and concentrated as described elsewhere.[1] All toxin analysis results in this study are reported as mass of toxin per mass of freeze-dried biomass. A potential source of variability arises if all samples are not dried to the same low residual water content. To study the possible differences in dry weight of freeze-dried samples, twenty biomass samples were divided and freeze-dried on two different types of freeze-dryers. After the freeze-drying, all forty samples were heated at 105°C for 12 h and then were re-weighed to

determine any loss due to residual water. The use of two
types of lyophilizers showed that the residual water levels
can vary by up to 10% and thus affect the toxin levels which
are reported on a dry weight basis. When samples were frozen
as collected and placed in a freeze-drying chamber type of
lyophilizer, the residual water contents were 10.9% on
average (n=20). When samples were shell frozen and connected
externally to a lyophilizing chamber, the residual water
content was found to be 3.8% on average (n=20). The combined
average residual water content was 7.3% (coefficient of
variation (CoV)=60.2%, n=40).

Sources of variability in the extraction and analytical
procedures were also investigated. The HPLC analytical
method of Harada et al.[2] was modified slightly, as described
by Kenefick et al.[3], for our work. The majority of the
microcystins present in bloom biomass have been reported[4] to
be extracted after two sequential extractions when using a 5%
acetic acid solution. The first extraction recovers
approximately 75% of the total toxin that can be extracted
with two extractions. Our studies showed the first 5% acetic
acid extraction recovered from 79% to 87% of the total toxin
recovered from two extractions. The precision of the 5%
acetic acid extraction on subsamples of the one algal biomass
sample, was also investigated. For six replicate
extractions, the toxin levels varied from 619 µg/g to 678
µg/g (CoV=3.7%, n=6). The analytical precision was studied
using a standard solution containing 10 mg/L microcystin LR.
The area counts from the microcytin LR peak showed a CoV=3.8%
(n=10).

The phosphatase assay for the quantitation of trace
levels of microcystins in water was performed as described by
Lambert et al.[5]. The CoV for triplicate analyses on 63
individual samples ranged from 1% to 47%, with a mean CoV=18%
(n=33) on raw water and a mean CoV=27% (n=30) on treated
water.

The variability in sampling was seen as a major factor
in the reported toxin concentrations in blooms and affected
waters. An intensive sampling program to evaluate spatial and
temporal variablity will be reported in detail elsewhere[6],
but the general findings are that these variations contribute
much greater uncertainty than any factors within the control
of the analyst. In particular, spatial variation yielded a
CoV=36% (n=29) in toxin levels found in biomass for one lake
and CoV=186% (n=45) for another lake. Temporal variation
over a 24 hour period produce a CoV=48% (n=18). Multiple
bloom samples collected from the same sampling point at the
same time produced a CoV=9% (n=5).

For microcystins in water, as measured by the
phosphatase assay, the CoV for hourly fluctuations over an
11.5 hour period for raw and treated waters were CoV=59%
(n=10) and CoV=10% (n=9), respectively. The CoV for the
daily fluctuations of single samples of raw and treated

waters over a 5 week period were CoV=71% (n=15) and CoV=20% (n=15), respectively.

3. RISK CHARACTERIZATION - SOURCES OF UNCERTAINTY

The prevalence of microcystin LR producing cyanobacterial species and the severity of potential health risks indicate the need for reliable and consistent monitoring and reporting methods so the public health significance can be more thoroughly understood.

As with most environmental health issues, the largest remaining source of uncertainty in the overall determination of health risk resides with the uncertain aspects of the toxicology (toxicokinetics in humans, character of human health effects and nature of dose-response relationships). We have demonstrated that there can also be substantial uncertainty contributed by the sampling component of the exposure assessment.

The contribution to overall uncertainty from analytical methodology can also be critical unless accurate and reliable techniques are carefully used. However, we conclude that is possible to reduce analytical uncertainty to negligible levels relative to the other dominant sources of uncertainty in characterizing potential health risks of microcystin occurrence in drinking water supplies.

ACKNOWLEDGEMENTS

This research was funded by a Strategic Grant from the Natural Sciences and Engineering Research Council of Canada.

REFERENCES

1. B.G.Kotak,S.L.Kenefick, D.L.Fritz, C.G.Rousseaux, E.E.Prepas and S.E. Hrudey, Water Res.. 1993, 27, 495.
2. K.I.Harada, K.Matsuura and M.Suzuki J. Chromatog., 1988 448, 275.
3. S.L.Kenefick, S.E.Hrudey, E.E.Prepas, N.Motkosky and H.Peterson Water Sci.Technol., 1992, 25, 2, 147.
4. D.J.Flett and B.C.Nicholson Toxic Cyanobacteria in Water Supplies: Analytical Techniques. 1991 Urban Water Research Association of Australia Research Report No. 26.
5. T.W.Lambert, M.P.Boland, C.F.B.Holmes and S.E.Hrudey, Environ Sci Technol., submitted and in review.
6. B.G.Kotak, A.Lam, E.E.Prepas, S.E.Hrudey and S.L.Kenefick, in preparation for submission to Toxicon.

Standing Committee of Analysts

D. Westwood

DEPARTMENT OF THE ENVIRONMENT, DRINKING WATER INSPECTORATE,
ROMNEY HOUSE, 43 MARSHAM STREET, LONDON SW1P 3PY, UK

For the benefit of those members of the audience who are not familiar with the work of the Standing Committee of Analysts a brief introduction of its history will be given. The SCA is a continuation of former Government Standing Committees and its background can be traced to the publication of the 1904 report of the Royal Commission on Sewage, through the 1929 compendium on Methods of Chemical Analysis as applied to sewage and sewage effluents, right up to the 1972 Green Book, 'The Analysis of Raw, Potable and Wastewaters'. The SCA was set up in 1972 by the Secretary of State as 'The Standing Committee of Analysts to Review Methods for Quality Control of the Water Cycle'. The principal objective was to produce recommended methods of analysis that are accurate, precise and reliable.

The terms of reference of the committee are to continuously review recommended methods for the analysis of water, sewage, effluents and associated sludges and sediments; and to update and provide new methods as necessary; it has also to advise on the needs and priorities for research in these areas.

The present structure of the SCA is comprised of a main committee, some nine working groups, each working in specific areas, ranging from sampling and accuracy, metals, microbiology, physical and empirical to radiochemical methods, and approximately 25 panels each reporting to individual working groups. The working group responsible for algal toxin analysis is the Organic Impurities working group and the panel consists of approximately 20 members from a wide variety of backgrounds with an interest in algal toxin analysis.

To give an idea of the size of SCA, there are approximately 30 members on the main committee which steers SCA and approximately 350 other members on the working groups and panels, including about 25 on the algal toxin panel. In general, panels and working groups are responsible for drafting and developing the methods, which after approval by the main committee, are published and distributed through Her Majesty's Stationery Office. To date, over 150 publications or Bluebooks, as these methods are commonly referred to, have been published. Each publication may contain several methods. One of the latest publications is an index which lists all parameters and some of the more important topic references. Most of you are aware that the SCA is now managed by the Drinking Water Inspectorate, although it is important to note that SCA covers the whole of the water cycle and not just drinking water.

Previous speakers have dealt with various methodologies used for the analysis of algal toxins and in theory all these could be used in SCA methods. Indeed nearly all the techniques explained have been used at some stage by the committee in production of bluebook methods. Any method, by any technique could be considered; the only provision being that no particular product, service or item of equipment will be endorsed by the committee.

The criteria for the production of methods can be briefly described in the following way. First of all, a need for the method must be demonstrated. This can be shown in any number of ways: (a) there may well be legislative requirements for a method, for example, analysis of the red list substances prescribed by regulations; regulations include, for example requirements to monitor pesticides; (b) it may be necessary to anticipate future analytical requirements, possibly for statutory purposes; in this case bromate is a good example; (c) it may be developed to support a monitoring strategy which seeks to investigate the significance of a potential problem; a good example here is algal toxin analysis.

The methods considered also need to reflect the different types of matrix, for example river waters, effluents, drinking waters and so on. Methods must also include an appraisal of potential interferences from the components of different matrices. The availability of laboratory equipment, expertise and experience should also be taken into account when recommending methods, although use of advanced instrumentation should not preclude a method simply because of the restricted availability of that specific piece of equipment. SCA publications are written in a format which facilitates adaptation of a method to suit a particular laboratory's requirement. Not all laboratories have the same equipment, or indeed the staff and experience. The number and type of samples requiring analysis will differ, as will the range of potential interferences. However the same basic technique can be used. Bluebook methods are therefore written so that they can be adapted to suit individual laboratory needs whilst retaining the fundamental principles of the analysis.

A fundamental requirement of all bluebook methods is that the analytical performance must be evaluated and reported. There are generally four categories of publication depending on the extent of performance. A recommended method is one that has been tested in 4-5 laboratories and has been shown to be satisfactory. A tentative method is one that has undergone testing in only 1-3 laboratories. A note to a method generally means that the method, although evaluated, has not been thoroughly tested, but that publication is required. Finally, a provisional method is one where some degree of testing has been carried out, but the robustness of the technique has not been conclusively demonstrated or it is likely that it will be replaced when an alternative method is available.

The analysis of algal toxins is an area where it is particularly important to demonstrate reliable performance as significant expenditure may be incurred as a result of that analysis. This expenditure might be in the form of treatment costs, loss of revenue from recreational use of water, or indeed in the possible costs of abortive analysis. For these reasons, analysts must be confident that the performance is acceptable, in terms of limits of detection and deviation, accuracy and precision. This performance in turn is dictated by the use which will be made of the analytical data.

SCA methods are recommended by the Drinking Water Inspectorate as being capable of achieving the necessary performance in terms of limits of detection and deviation needed to satisfy the requirements of the Water Supply (Water Quality) Regulations 1989. The Guidance Document which supports the Regulations advises that the total error must not exceed 20% and the limit of detection should be at least 10% of the level of interest. Certainly with the analytical techniques available to us at present, these requirements will be difficult to meet for many fields of trace organic analysis, unless of course new technologies are developed. It is here that the users of SCA methods, with information on the analysis of algal toxins can help. Ideally, for methods to be published by SCA, results from several laboratories are necessary to assess performance. It would be helpful if users could provide information about the performance of the methods they are using. In terms of water supplies, the Drinking Water Inspectorate uses SCA methods as a yard stick against which other methods are assessed and information from users would make this task that much easier.

In conclusion, it must be emphasised the importance given to ensuring that the analysis of algal toxins is carried out using methods that are robust and proven in terms of their accuracy, precision and limit of detection.

References

1   Water Supply (Water Quality) Regulations - Statutory Instrument 1989 No. 1147, as amended by SI 1989/1384, SI 1991/1837, SI 1991/2790.

2   Guidance Document: "Guidance on Safeguarding the Quality of Public Water Supplies" HMSO London ISBN 0117522627.

3   Index of Methods for the Examination of Waters and Associated Materials 1976-1992. HMSO (ISBN 011 752669 X).

Poster Presentations

Toxic Cyanobacteria (Blue-green Algae) in Portuguese Freshwaters

V. Vasconcelos

INSTITUTO DE ZOOLOGIA "DR. AUGUSTO NOBRE", FACULDADE DE CIÊNCIAS, PRAÇA GOMES TEIXEIRA, P-4000 PORTO, PORTUGAL

1 INTRODUCTION

Human health problems related to toxic cyanobacteria are common all over the world and the research on the effects of their toxins has been developing especially after the reports on tumour promotion properties of the microcystins[1,2].

In Portugal, potentially toxic cyanobacteria are common in natural lakes, reservoirs and also in large rivers[3-8], although data on toxicity is scarce[9-11]. Although there are not many reports on human intoxications due to cyanobacteria toxins in Portugal[9], the occurrence of blooms in water supplies used for consumption and recreation led us to suspect that human intoxications are more frequent than those cases reported. In this paper, data on the occurrence of toxic cyanobacteria blooms in Portuguese freshwaters are presented.

2 MATERIAL AND METHODS

An analysis of Portuguese literature concerning freshwater phytoplankton was undertaken in order to select the most eutrophic waterbodies. Data from 61 lakes, reservoirs and rivers were analysed and 36 of them were sampled from 1989 to 1992. Some of the sites were sampled several times during this period. Blooms were collected and bioassayed. Toxicity was measured by intraperitoneal (i.p.) mouse bioassay using 20-30 g male Charles River mice (I. Gulbenkian, Oeiras). Suspensions of freeze-dried cells were injected in pairs of mice per dose level. Symptoms were registered and dead animals were observed internally for signs of hepatotoxicity. Cyanobacteria were considered toxic if death occurred at doses < 1500 mg/kg. LD50 was determined as the dose between those that produced 0% and 100% mortality.

3 RESULTS AND DISCUSSION

Thirty bloom samples were analysed and 60% of them were found to be toxic by mouse bioassay. All the toxic bloom samples were hepatotoxic and death occurred between 30 minutes and 2 hours. Livers were enlarged and dark red weighting 8% to 10% of total body weight.

The occurrence of the dominant cyanobacteria species in toxic blooms is shown in table 1. Other cyanobacteria species were A. spiroides, Aphanizomenon flos-aquae, Oscillatoria tenuis and Gomphosphaeria lacustris.

LD50 values of the blooms ranged from 20 mg/kg to 700 mg/kg. The main part of the blooms showed high toxicity levels, with over 60% of them with an LD50 < 100mg/kg.

Toxic blooms were distributed across the country. Concentrations were found in the river Minho and river Douro in the north, lakes of the Quiaios and Mira lake area in the central area, the Alentejo and Algarve reservoirs and Guadiana river in the south. It is known that cyanobacteria are also common in the Azores islands[12], although no data on toxicity is available as yet.

Toxicity was found to vary considerably with time within a lake. In Mira lake in central Portugal, toxicity of bloom material collected in 1992 and composed mainly of M. aeruginosa and An. flos-aquae, varied in toxicity between lethal doses of 37.5 mg/kg and 400 mg/kg between June 5 and June 11, yet in October and November the blooms were nontoxic at doses of up to 1500 mg/kg.

In Minho estuary, clams and mussels grow naturally and are harvested and sold in local markets. Knowing that molluscs may accumulate cyanobacterial peptide toxins[13,14] and that toxic Microcystis blooms are common in this river, we predict that human health hazards may occur there.

In Guadiana river an episode of human intoxication occurred in 1987. Several people suffered dermatitis and gastrointestinal disorders after drinking water from a heavy bloom of Aphanizomenon flos-aquae[9]. In 1992, at the same site, a M. aeruginosa/M. wesenbergii bloom showed an LD50 of 35 mg/kg. This river is still used as a drinking water source.

These results show that toxic cyanobacteria blooms are common in Portuguese freshwaters.

Table 1 - Occurrence of the dominant cyanobacteria species in toxic blooms collected during 1989-1992 in Portuguese freshwaters (frequency/total)

SPECIES	FREQUENCY/TOTAL
Microcystis aeruginosa	13/18
M. wesenbergii	1/18
Anabaena flos-aquae	5/18
Nostoc sp.	1/18

Most of the sampled sites are used for recreation or as drinking water sources, which lead us to recommend that there should be a monitoring programme established for these waterbodies during summer months in order to detect and quantify cyanobacterial toxins.

In most of the portuguese water treatment plants, only chlorination and filtration are used and these treatments are not effective for the removal or destruction of the toxins. Bioassays, imunoassays or chemical techniques such as High Pressure Liquid Chromatography (HPLC) or Thin Layer Chromatography (TLC) should be used whenever the amount of cyanobacteria in the raw water reaches high values.

As toxicity may vary considerably within a lake over a few days it is recommended that blooms are considered toxic whenever they occur and direct contact with the cyanobacteria should be avoided.

4 REFERENCES

1. I.R. Falconer, Env. Toxicol. Water Quality, 1991, 6, 177
2. R. Nishiwaki-Matsushima, T. Ohta, S. Nishiwaki, M. Suganuma, K. Koyama, T. Ishikawa, W.W. Carmichael, H. Fujiki, J. Cancer Res. Clinical Oncol., 1992, 118, 420
3. G. Cabeçadas, M.J. Brogueira, J. Windorf, Int. Revue ges. Hydrobiol, 1986, 71, 795
4. M.R.L. Oliveira, Bol. Inst. Nac. Invest. Pescas, 1984a, 11, 3
5. M.R.L. Oliveira, Bol. Inst. Nac. Invest. Pescas, 1984b, 11, 45
6. M.R.L. Oliveira, Bol. Inst. Nac. Invest. Pescas, 1984c, 12, 37
7. M.M.C. Silva, Pub. Inst. Zool. "Dr. A. Nobre", 1989, 215, 1
8. V.M. Vasconcelos, Arch. Hydrobiol., 1991, 121, 67
9. M.R.L. Oliveira, Relat. Téc. Cient. INIP, 1991, 42, 1
10. V.M. Vasconcelos, Verh. Internat. Verein. Limnol, 1993, 25, 694
11. V.M. Vasconcelos, W. Evans, W.W. Carmichael, M. Namikoshi, J. Env. Sci. Health, 1993, 28A(9), 2081
12. M.C.R. Santos, A.M.F.R. Rodrigues, P. Sobral, F.J.P. Santana, Actas da 3ª Conferência Nacional sobre a Qualidade do Ambiente, 1992, A.R. Pires, C. Pio, C. Boia, T. Nogueira (eds.), Aveiro, 217
13. J.E. Eriksson, J.A.O. Meriluoto, T. Lindholm, Hydrobiologia, 1989, 183, 211
14. I.R. Falconer, A. Choice, W. Hosja, Env. Toxicol. Water Quality: An Int. Journal, 1992, 7, 119

Screening of Cyanobacterial Toxins in *Microcystis Aeruginosa* Collected from Blooms and Cultures

F. van Hoof,[1] T. van Es,[1] D. D'hont,[2] and N. De Pauw[2]

[1]STUDY CENTER FOR WATER, MECHELSESTEENWEG 64, B-2018 ANTWERPEN, BELGIUM

[2]LABORATORY FOR BIOLOGICAL RESEARCH IN AQUATIC POLLUTION, UNIVERSITY OF GENT, J. PLATEAUSTRAAT 22, B-9000 GENT, BELGIUM

1 INTRODUCTION

Following the detection of toxic Microcystis aeruginosa blooms in several surface water storage reservoirs through in vivo toxicity testing, an investigation was started in order to find more rapid alternatives to bioassays for the detection of cyanobacterial toxins in algae collected from blooms. Three isolation methods (extraction with aqueous acetic acid[1], ultrasonic extraction with water and extraction with a mixture of water-methanol-butanol[2]) were tested. The extracts were analysed using reversed phase HPLC with UV detection .The method selected was used to analyse samples collected from storage reservoirs used for drinking water production and samples grown under laboratory culture conditions.
LC-Thermospray Mass Spectrometry and Fast Atom Bombardment Mass Spectrometry have been used in order to obtain information on the identity of the toxins .

2 EXPERIMENTAL

In most experiments algal samples equivalent to 0.02 gram dry weight were handled. In the first method (extraction with aqueous acetic acid) the sample was extracted three times with 5 ml 5% acetic acid by shaking for five minutes. The combined extract was centrifuged and passed over a C_{18} cartridge (SepPak Millipore) which had been preconditioned with 10ml methanol , followed by 15ml water:methanol(4:1).
Elution was performed with 2ml methanol. The extract was evaporated to dryness, dissolved in 1ml acetonitrile and analysed.
In the second method extraction was done with ultra pure water (100μl/mg) through ultrasonication for 5 minutes at 30-40°C. After centrifugation the extract was filtered over a 0.45 μm filter, dried and taken up in the mobile phase and analysed.
In the third method the extraction was carried out through ultrasonication for 5 minutes at 30-40°C with a mixture water:methanol:butanol(75:20:5). The extracts were combined , dried , taken up in the mobile phase and analysed.

For chromatographic analysis two systems were used :
 a 25cm; 4.6mm I.D.C_{18} column (Baker) ; particle diameter:
5 µm and an internal surface reversed phase (ISRP) column
combining both size exclusion and reversed phase chromato-
graphy.The column used was an ISRP 130 C_{18} column(Société
Française des Colonnes),15 cm long,5µm particle diameter.In
both cases the mobile phase was 26% acetonitrile-74% 0.01M
ammonium acetate at a flow of 1.0 ml/min . UV detection at
240nm ;injection volume:50µl. Cyanazine was used as an
internal standard for the calculation of relative retention
times.
The reversed phase C_{18} column was used in most of the work
since initial experiments pointed out that it was more
efficient in separating the peaks of interest. Using the same
chromatographic conditions,structural information on the
compounds detected was sought through LC-Thermospray Mass
Spectrometry.Some chromatographic fractions were analysed
using Fast Atom Bombardment Mass Spectrometry (FAB-MS) .
A Microcystis aeruginosa sample from Lake Akersvatn , Norway
(obtained from O.M.Skulberg,Norwegian Institute for Water
Research), in which microcystin-LR had been detected, was used
as a positive control throughout the study.
Microcystis aeruginosa (obtained from Prof.L.Mur,University of
Amsterdam)was grown in BG11 medium at different temperatures
(21,24,25 and 29°C) and different light intensities(1000-4000-
8000 lux) in order assess the influence of environmental
conditions on toxin production.
Bioassays were carried out through intraperitoneal injection
in BALB/c albino mice.

3 RESULTS AND DISCUSSION

The extraction methods were evaluated on several samples which
had been proven toxic through mouse bioassays.The extraction
technique using 5% acetic acid produced less chromatographic
information than both other extraction techniques.These gave
similar results for both the number of peaks detected and for
their intensity.Therefore the ultrasonication technique using
water as extractant was selected for further experiments.The
concentration-purification step over C_{18} columns was evaluated
separately in order to exclude losses during this step.
No differences were observed in the intensities of the peaks
of interest before and after C_{18} enrichment.
All samples from toxic Microcystis aeruginosa blooms in reservoirs and from Lake Akersvatn showed at least two distinct
peaks in the chromatograms.The latter sample showed major
peaks with relative retention times towards cyanazine of
0.30; 0.50; 0.84 and 0.90. The peaks in the chromatograms from
surface water storage reservoirs had different retention times
from the Lake Akersvatn sample:0.38; 0.52 and 0.70 respecti-
vely.None of the Microcystis aeruginosa samples grown under
laboratory conditions revealed any peaks in the chromatograms when extractions were performed on the same amount of
algae on a dry weight basis.Minor peaks could only be detected

after analysing 100mg samples on dry weight basis and after concentration of the samples to 500µl.The relative retention times of the peaks detected were: 0.26; 0.33; 0.42; 0.47 and 0.49.The laboratory grown Microcystis aeruginosa samples did not reveal any toxicity after intraperitoneal injection in BALB/c albino mice.

Information on the identity of the peaks was sought through LC-Thermospray Mass Spectrometry.For this purpose two grams of algae were extracted. The extract was finally taken up in 2ml of the mobile phase. In the sample of Lake Akersvatn the peak with relative retention time 0.30 gave a mass spectrum with a pseudomolecular ion of 995,indicating a molecular ion of 994.This corresponds exactly to the toxin described for this Microcystis aeruginosa sample by Berg et al.[3] The peaks in the chromatograms of the toxic Microcystis aeruginosa samples from the storage reservoirs investigated did not reveal this component .Fast Atom Bombardment Mass Spectroscopy applied on these samples gave indications for the presence of compounds with molecular weight of 786(peak with relative retention time 0.52) and 1129(peak with relative retention time 0.70).

4 CONCLUSIONS

The extraction technique using ultrasonic extraction with water was as efficient as the technique using a mixture of methanol,butanol and water as extractant and was superior to extraction with aqueous acetic acid .

Using the same amount of algae and applying the same concentration factors,only toxic Microcystis aeruginosa samples showed distinct peaks in the chromatograms .

Using the extraction technique described above and LC-Thermospray MS,the presence of microcystin-LR could be confirmed in a sample from Lake Akersvatn,Norway.

The presence of microcstin-LR could not be confirmed using the mass spectrometric techniques mentioned above in toxic samples from one of our storage reservoirs(Broechem).

5 ACKNOWLEDGEMENT

LC Thermospray Mass Spectrometry and Fast Atom Bombardment of LC fractions were carried out by dr.W.Lauwers,Janssen Pharmaceutica,Beerse,Belgium.

6 REFERENCES

1.M.F.Watanabe,K.I.Harada,K.Matsura,M.Watanabe and M.Suzuki, J.Appl.Phycol.,1989,1,161.
2.J.A.O.Meriluoto and J.E.Eriksson,J.Chromatography,1988, 438 ,93.
3.K.Berg,W.W.Carmichael,O.M.Skulberg,C.Benestad and B.Underdal,Hydrobiologia,1987,144,97.

Toxicity Studies with Blue-green Algae from Flemish Reservoirs

F. van Hoof,[1] Ph. Castelain,[2] M. Kirsch-Volders,[2] and
J. Vankerkom[3]

[1]STUDY CENTER FOR WATER, MECHELSESTEENWEG 64, B-2018 ANTWERPEN,
BELGIUM

[2]LABORATORIUM VOOR ANTROPOGENETICA, VRIJE UNIVERSITEIT BRUSSEL,
PLEINLAAN 2, B-1050 BRUSSEL, BELGIUM

[3]VLAAMSE INSTELLING VOOR TECHNOLOGISCH ONDERZOEK, BOERETANG 200,
B-2400 MOL, BELGIUM

1 INTRODUCTION

The occurence and the effects of blue-green algal blooms has
been intensively studied during the last decade[1-3] .
Although blue-green algal blooms occur frequently in Flemish
reservoirs used for the production of drinking water,no
information was available on their eventual toxic effects.
Therefore an investigation was started in which samples
collected from different reservoirs,including Microcystis
aeruginosa,Anabaena flos-aquae,Aphanizomenon flos-aquae and
Oscillatoria agardhii,were tested for acute toxicity using
animal bioassays.The most toxic species were selected for
skin irritation,skin sensitisation and mutagenic effects in
the Salmonella typhimurium test .

2 EXPERIMENTAL

All samples collected were kept at -25°C before testing .
After defrosting,the samples were centrifuged at 6000 rpm
for 15 minutes.The supernatant was used for tests on acute
toxicity,skin irritation and for skin sensitisation
studies.The freeze dried Microcystis aeruginosa sample from
Lake Akersvatn, Norway,which was obtained from Dr.O.Skulberg
(NIVA, Oslo)was suspended in 0.9%NaCl and was thereafter
treated in the same way as the other samples.
The samples used in the mutagenicity tests were freeze dried
and resuspended in 0.9% NaCl.The suspensions were centrifuged at 5000 rpm (20 minutes) and 15000 rpm (15 minutes),and
then filtered successively over 5 μm ; 1.2μm; 0.45μm and 0.2
μm filters.
Acute toxicity tests were performed using three months old
BALB/c albino mice.The supernatant of the algae is administered by a single intraperitoneal injection. For each
sample the experiments were carried out at minimum five dose
levels. The LD_{50} values were calculated using the method of
Marquardt[4] .Clinical observations of the animals are made

frequently during the first days and daily over a period of
14 days maximum.
Acute skin irritation was evaluated by application of 0.5 ml
liquid on the shaved, intact skin of New Zealand White
rabbits for 4 hours. Observations were made 24,48 and 72
hours after patch removal. Three rabbits were used in each of
the tests.
Skin sensitisation was examined by applying 0.3 ml liquid
during 6 hours on the intact skin of Hartley guinea pigs.
Symptoms for delayed contact hypersensitivity were scored 24
and 48 hours after the exposure was stopped. In each test 20
exposed and 10 control animals were used.
Mutagenic activity was tested in the <u>Salmonella typhimurium</u>
assay on five strains: TA1537, TA1538 and TA98, for the
detection of frameshift mutagens and TA1535 and TA100, for
the detection of base pair substitution mutagens. Each sample
was tested at four different concentrations with and without
enzyme activation. The Arochlor 1254 induced 9000g fraction
of the livers of male Sprague-Dawley rats was used for
enzyme activation. NaN$_3$(TA1535), 9-aminoacridine (TA1537) and
4-nitroquinoline-N-oxide (TA1538, TA98, TA100) were used as
positive controls in the tests without enzyme activation; 2-
aminoanthracene (TA1535) and 2-aminofluorene (TA1537, TA1538,
TA98, TA100) were used for the same purpose in the tests with
enzyme activation.

3 RESULTS AND DISCUSSION

The results of the acute toxicity tests are summarised in
Table 1. All LD$_{50}$ values are expressed in mg dry weight/kg
body weight.

<u>Table 1</u> Results of the acute toxicity tests

Species	Code	LD$_{50}$
Microcystis aeruginosa	7A	83
Microcystis aeruginosa	7B	701
Microcystis aeruginosa	7E	259
Microcystis aeruginosa	7F	99
Microcystis aeruginosa	7C	80
Microcystis aeruginosa	MBXX	85
Microcystis aeruginosa	MBR	187
Microcystis aeruginosa	MBII	56
Microcystis aeruginosa	MBIV	66
Microcystis aeruginosa	MADD	53
Aphanizomenon flos-aquae	10A	>500
Aphanizomenon flos-aquae	10B	>500
Aphanizomenon flos-aquae	APXX	270
Oscillatoria agardhii	3	>890
Oscillatoria agardhii	5	>1150
Oscillatoria agardhii	6	>1060
Oscillatoria agardhii	1B	>780
Anabaena flos-aquae	ABXX	>150

All Microcystis samples produce toxic effects after intraperitoneal injection.All the other samples,except one sample of Aphanizomenon flos-aquae,do not induce mortality. An inverse relationship was observed between the dose injected and the survival time.The highest doses injected resulted in death within two hours,except for sample MBXX, for which death occured between 2-18 hours.After injection of sample MADD (Lake Akersvatn) all animals died within one hour.The most toxic Microcystis samples collected produce LD$_{50}$ values similar to Microcystis sample collected from Lake Akersvatn.An LD$_0$ for this sample of 8-15 mg/kg has previously been reported[5].No lethality was observed after dosing two of the more toxic samples(7F and MBII) orally up to 250 mg/kg .
Four of the acutely toxic samples(MBII,MADD,MBXX and 7C) were tested for skin irritation using New Zealand White strain rabbits. None of them produced any adverse effects. Two of these samples(7C and MBII)were tested for skin sensitisation using Hartley guinea pigs.No symptoms of skin sensitisation were found.
Three samples[7C,MADD and APXX(Aphanizomenon-flos aquae)] were tested in the Salmonella typhimurium mutagenicity assay using five different strains.The Microcystis 7C sample shows a doubling of the spontaneous mutation frequency at a concentration of 25µg/plate in strain TA1535 without metabolic activation. At higher concentrations however the number of revertants decreases without indications for toxic effects.A dose dependent increase,leading to a doubling of the spontaneous mutation frequency,was seen in the same strain with and without metabolic activation after applying the Aphanizomenon sample.This effect could not be reproduced in further experiments .No increases in gene mutations were observed in any of the other strains .

4 REFERENCES

1.O.M.Skulberg,G.A.Codd and W.W.Carmichael,Ambio,1984,14,244
2.W.W.Carmichael,C.L.A.Jones,N.A.Mahmood and W.C.Theiss,CRC Critical Reviews in Environmental Control,1985,15,275.
3.National Rivers Authority,Water Quality Series No.2, 1990:Toxic Blue-Green Algae.
4.D.W.Marquardt,J.Soc.Ind.Appl.Math.,1963,2,431.
5.K.Berg,W.W.Carmichael,O.M.Skulberg,C.Benestad and B.Underdal,Hydrobiologia,1987,144,97.

Cases of Cyanobacterial Toxicoses on Swiss Alpine Pastures

K. Mez,[1] H.-R. Preisig,[1] B. Winkenbach,[1] K. Hanselmann,[1] R. Bachofen,[1] B. Hauser,[2] and H. Nägeli[2]

[1]INSTITUTES OF PLANT BIOLOGY AND SYSTEMATIC BOTANY, UNIVERSITY OF ZÜRICH, SWITZERLAND

[2]INSTITUTES OF PATHOLOGY AND PHARMACOLOGY/TOXICOLOGY, UNIVERSITY OF ZÜRICH–TIERSPITAL, SWITZERLAND

1 INTRODUCTION

During the last two decades about 100 unexplainable deaths of cows have been reported from the South-Eastern Alps in Switzerland. All the deaths occurred on alpine pastures at an altitude of 2100–2300 m and most of them after long warm and dry periods in August or September. Based on the results of the histopathological examinations of the dead animals we follow the hypothesis that the cows died of cyanobacterial (blue–green algal) toxicoses.

In 1991 we selected two areas for detailed investigation of the algal composition, the toxicity of the algae and the ecological background of the alpine aquatic ecosystems. The two experimental areas are situated at 2300 m on silicate rock. Area 1 ("Confin") consists of several ponds and a bog which are fed by rain and snow-melt water ("A" in Table 1) and a small brook which bears spring water containing relatively high amounts of minerals ("B" in Table 1). In 1990 ten cows had died at Confin. Area 2 ("Tamboalp") is situated below a glacier lake and includes the glacier river ("C" in Table 1), a bog and two small moraine ponds ("D" in Table 1) fed by snow and rain. In this area four cows had been found dead in August 1992.

During the last decade toxin(s) originating from Veratrum album (False Helleborine) and from fungi as well as lithological substances and run-off from military waste could be excluded as sources of poisoning. We therefore investigate the hypothesis that the cows died of cyanobacterial hepatotoxins.

2 PATHOLOGY

The cows died within a short time after showing restlessness, tremor, spasms, staggering and lowing. The histopathological examination of the dead animals revealed severe changes in the liver tissue. The centrilobular parts of the liver were blood-filled whereas the periphery was totally pale. The colour of the centrilobular hepatocytes' cytoplasm was heterogeneously granular and the chromatin condensed or even destroyed.

<u>Table 1</u> Water temperature (diurnal changes in $^\circ$C), average conductivity (μS/cm), pH, alkalinity (french degrees) and nutrient composition (μM) of the ponds (A) and the river (B) at Confin, of the glacier river (C) and the moraine pond (D) at Tamboalp and of Lake Zürich[4] (E, at 30 m).

	A	B	C	D	E
water temperature	5–30	3	8–11	n.d.	n.d.
conductivity	8–65	115	30–60	6–10	n.d.
pH	5–7.7	7.5	7–8.5	n.d.	7.7
alkalinity	n.d.	n.d.	4	1.5	13.8
ammonia	<45	3	<0.9	<0.9	<7
phosphate	<2	<0.3	<0.3	<0.3	245
calcium	10–200	400	200	35	1200
magnesium	4–40	130	30	3	250
sodium	20–150	100	20–90	17	80
potassium	5–60	15	10–18	5	n.d.
sulfate	11–80	200	60–65	n.d.	160
nitrate	1–20	20	20–50	n.d.	55
chloride	<70	<30	<30	n.d.	80

n.d.   not determined

3 ALGAL COMPOSITION AND TOXICITY TESTS

Cyanobacteria commonly occur at both sites (e.g. <u>Merismopedia, Oscillatoria, Pseudanabaena, Scytonema, Scytonematopsis, Synechococcus</u> etc.). Except for a bloom of the green flagellate <u>Chlamydomonas botryopara</u> at Confin, no algal blooms could be observed. However a striking mass development of a benthic cyanobacterium, <u>Phormidium autumnale</u>, occurred at Tamboalp. In the fast-running muddy water of the glacier river this organism formed large mats on the ground and even underneath stones.

Toxic cyanobacteria have not been found until now. <u>Scytonematopsis starmachii</u> and <u>Phormidium autumnale</u> were tested for toxicity by mouse bioassay (courtesy of G. Codd, University of Dundee) and for future investigations we recently established a protein phosphatase inhibition test, using phosphatases extracted from oilseed-rape seeds and [^{32}P]phosphorylase a as a substrate[1-3].

4 PHYSICO-CHEMICAL CHARACTERISTICS OF THE ALPINE AQUATIC SYSTEMS

All organisms living in the respective alpine waters are confronted with a very low nutrient content. Table 1 shows the physico-chemical background of the examined waters at Confin and Tamboalp and, for comparison, of Lake Zürich[4].

Due to biological activity (i.e. photosynthesis and respiration), conductivity, pH and oxygen concentration in the water fluctuate in diurnal cycles. On sunny days, solar irradiance is so high that it inhibits the photosynthetic activity of the algae (photoinhibition). Rain water influences the chemical composition of the water by the quantity and quality of its constituents. Conductivity can double due to atmospheric wet depositions.

5 CONCLUSIONS

The organisms in the examined ecosystems are challenged by several extreme situations:
- The ponds and brooks are highly oligotrophic, the nutritional conditions are comparable to those at the end of an algal bloom: nutrients are limited and/or unbalanced and this stress might eventually induce toxin production. It is interesting that some of the species found in abundance (e.g. *Phormidium* *autumnale*) are benthic organisms. They are to be examined for their capacity to scavenge nutrients adsorbed to the surface of sediment particles.
- Alpine aquatic ecosystems are exposed to very high solar irradiance. Whether light stress could stimulate toxin production in cyanobacteria is of great interest with respect to possible climatological changes (i.e. ozone depletion).
- Due to the low alkalinity of the water, the aquatic systems are very sensitive to changes in the ion composition, especially the proton concentration. Acid rain leads to changes in conductivity and pH of the pond water; their influence on species composition and toxin production needs to be studied.

Although the alpine aquatic ecosystems do not correspond to textbook descriptions of cyanobacterial habitats we follow the hypothesis that the above mentioned deaths of cows on alpine pastures are a product of cyanobacterial toxins. For our future field work we hope that the phosphatase inhibition test will enable us to detect even smallest amounts of hepatotoxins.

ACKNOWLEDGEMENTS

This research project is supported by a NFP31 grant of the Swiss National Research Foundation (No. 4031-033432.92) and the Swiss Federal Department for Veterinary Medicine (grant No. 012.91.11).

REFERENCES

1.   C. MacKintosh, K.A. Beattie, S. Klumpp, P. Cohen and G.A. Codd, FEBS lett., 1990, 264, 187
2.   C. MacKintosh and P. Cohen, Biochem. J., 1989, 262, 335
3.   P. Cohen, S. Alemany, B.A. Hemmings, T.J. Resink, P. Strålfos and H.Y. Lim Tung, Methods Enzymol., 1988, 159, 390
4   L. Sigg and W. Stumm, 'Aquatische Chemie', Verlag der Fach-vereine, Zürich, 1989

Biological and Economic Significance of Benthic Cyanobacteria in Two Scottish Highland Lochs

R. P. Owen

NORTH EAST RIVER PURIFICATION BOARD, GREYHOPE HOUSE, GREYHOPE
ROAD, TORRY, ABERDEEN AB1 3RD, UK

1 INTRODUCTION

Loch Insh (NGR NH832045) and Loch Morlich (NGR NH965094)
situated in the upper catchment of the River Spey, Highland
Region, Scotland are nationally important for nature conser-
vation, recreation and tourism.   These lochs are upland in
character, having glacial kettle-hole origins and receiving
drainage from extensive tracts of mountain and moorland,
although they are shallow relative to surface area (Table 1).

Table 1   Physical and Catchment Characteristics of Loch Insh
           and Loch Morlich

Physical Characteristics		
	LOCH INSH	LOCH MORLICH
Altitude (m)	220	319
Max. Length (km)	1.7	1.6
Max. Breadth (km)	1.1	1.0
Max. Depth (m)	44.0	12.5
Mean Depth (m)	11.4	6.5
Volume (Mm^3)	12.9	5.2
Area (Km^2)	1.1	1.3
Flushing Rate (days)	8.1	35.0

Catchment Characteristics		
	LOCH INSH	LOCH MORLICH
Mean Daily Flow (m^3/sec)	18.5	0.8
Catchment Area (km^2)	82.0	44.9
% Arable	0	0
% Forestry	5	20
% Mountain/Moorland/ Rough Grazing	95	80

Loch Insh forms part of an extensive Site of Special Scientific Interest while Loch Morlich lies within Glenmore Forest Park.   Both lochs accommodate water sports centres which offer a variety of activities, some with high risk of immersion, including canoeing, sail-boarding, and sailing. Bathing, paddling and shore-line walking are also commonly practised.   These lochs are major attractions, situated in superb mountain and forest scenery and the activities they support provide income and employment for this highland area. Any restriction on these recreational activities constitutes a threat to the local economy.

Visitors to these lochs have come to expect a pristine environment with very high water quality. In 1990, this reputation was tarnished by the deaths of three dogs being exercised at Loch Insh.   These animals had been observed to consume stranded, shoreline accumulations of algal material and to have died very shortly afterwards.   The shoreline material was found to consist very largely of the filamentous cyanobacterium, Oscillatoria. The cyanobacterial material was found to be highly neurotoxic by mouse bioassay with the alkaloid, anatoxin-a, being detected from samples of Oscillatoria and also from the stomach contents of one animal[1].

2 DESCRIPTION OF THE PROBLEM

At the time of the dog fatalities substantial accumulations of material were found in and out of the water margins at many locations on the windward shores of Loch Insh.   Adopting a precautionary approach, local Environmental Health officers posted signs warning visitors to avoid any scums or shoreline debris.   Not long after these events similar cyanobacterial material was discovered on the popular beach area of Loch Morlich and warning signs erected there too.   These cyano-bacterial events have recurred approximately from late July to early October annually at both lochs since 1990.   The general economic cost to local concerns from loss of business is not directly measurable though for the water sports operation at Loch Insh it is known to be significant.

The first obvious indications of cyanobacterial problems in these lochs in each of the last three years has been the sudden appearance of dark green masses of detached material driven onto wind exposed shores. Typically the quantity grad-ually increases to form stranded layers, up to 10 cm thickness in Loch Insh, with material also floating in a band up to two metres wide in the near shore areas.   The material varies in composition with several species of filamentous diatoms and green algae also present at first but with the proportion of Oscillatoria increasing until this genus dominates.   It is apparent that the algal/cyanobacterial admixture originates from benthic mats of these species growing profusely in the euphotic zone. Detachment of the mat is often associated with gas production and wind disturbance of the sediment.

3. TROPHIC STATUS

By normal means of classification both Loch Insh and Loch Morlich appear to be oligotrophic or weakly mesotrophic (mean annual concentrations not exceeding 3 ug/l Soluble Reactive Phosphorus; 8 ug/l Total Phosphorus and 9 ug/l Total Oxidised Nitrogen) and somewhat acidic (mean pH 6.1 in both). Open water chlorophyll concentrations are also very low and similar to previously recorded values[2]. In a local context, there are several lochs in the north east of Scotland with considerably greater nutrient concentrations but in which cyanobacterial productivity has remained negligible.

In the case of Loch Insh, a high throughput of water (Table 1) may preclude the formation of significant phytoplankton populations and favour the production of benthic mats in the euphotic zone. This scenario does not apply, however, in Loch Morlich which has a much lower water throughput although in this loch the production of benthic <u>Oscillatoria</u> is much less than in Loch Insh.

Although loch water chemistry indicates, in general, a paucity of nitrogen and phosphorus there are a number of factors to consider. Monitoring of these lochs has been regular but infrequent which may have lead to an underestimate of nutrient loadings, there have been few attempts to study possible sources of nitrogen and phosphorus in these catchments and the interaction between sediment chemistry and the production of <u>Oscillatoria</u> may be particularly important.

4. POSSIBLE SOURCES OF NUTRIENTS

Agricultural sources of nutrients have been shown to be important in eutrophication processes in other water bodies in the north east of Scotland[3]. However, in these two catchments there is negligible arable land[4] with the dominant land-uses being mature forest, rough grazing and grouse moor.

Treated sewage effluents are disposed of into both catchments though their significance and contribution to overall loadings of nitrogen and phosphorus are not yet known. There are four sewage treatment works, serving a combined resident population of approximately 4,000 people and considerably more summer visitors, discharging directly or indirectly to the River Spey no more than fifteen kilometres upstream of Loch Insh. At Loch Morlich there are several small units treating sewage from day lodges on the ski slopes of Coire Cas and Coire na Ciste and also sewage from the caravan site and village of Glenmore. All of this effluent discharges into the inflowing streams within a few kilometres of Loch Morlich. The population contributing sewage to these units also varies seasonally but, in this case, with the greater load being in the winter and early spring skiing season.

An added complication at Loch Insh is the extent of winter flooding which can increase the area of continuous standing water several fold compared to the summer surface area of the loch. This flooding has, in recent years, extended so far upstream that the sewage treatment works at Kingussie was inundated so that, essentially, untreated sewage escaped into the river. Furthermore, the periodic inundation of the Insh Marshes, an extensive wetland immediately upstream of the loch, may flush nutrients into the loch.

5. FURTHER RESEARCH

To assess the relative importance of catchment sources of nutrients, the development of Oscillatoria accumulations and the variation in toxicity of this material, a collaborative programme of research has commenced on Loch Insh. This programme is designed to gather chemical, biological and hydrological data on the loch and the catchment as well as the measurement of loads of nutrients from sewage treatment works. Work is also in progress by other groups on the nutritional requirements of Loch Insh Oscillatoria, sediment chemistry and relations with toxin levels.

6. CONCLUSIONS

1. The problems associated with nuisance production of cyano-bacteria are not confined to obviously eutrophic standing waters. Lochs Insh and Morlich appear to be oligotrophic or weakly mesotrophic and yet suffer regularly from accumulations of anatoxin-a producing, benthic Oscillatoria which has led to dog fatalities.

2. These lochs are major tourist attractions and support recreational activities providing local employment. The local economy of this highland area is threatened by these incidents.

3. Treated sewage effluents may be possible catchment sources of nutrients and loads from these may have been under-estimated. Agricultural sources are negligible. These are being investigated in a current research programme.

4. Benthic Oscillatoria may occur in other similar upland standing waters of apparently poor nutrient status and the extent of this phenomenon is as yet unknown.

References

1. C.Edwards et al., Toxicon 30, 1165-1175 (1992).
2. NERPB, Loch Morlich, A Baseline Study of the Catchment, Unpublished Report, 1981.
3. R.P. Owen, Causes and Effects of Nuisance Populations of Cyanobacteria in the Loch of Skene, Unpublished Report, NERPB, 1980.
4. W. Towers, Land Use in the River Spey Catchment, Symposium, Aberdeen Centre for Land-Use, University of Aberdeen, 1987.

First Results on the Occurrence of Microcystin-LR in Berlin and Brandenburg Lakes

Jutta Fastner

INSTITUTE FOR WATER, AIR AND SOIL HYGIENE, FEDERAL HEALTH OFFICE, CORENSPLATZ 1, D-14195 BERLIN, GERMANY

1 INTRODUCTION

Due to eutrophication water blooms of cyanobacteria are common in Berlin and Brandenburg water bodies. Many species of cyanobacteria are able to produce hepatotoxins (e.g. microcystin-LR) which can cause illness or death of animals or humans. Many cases of animal intoxication and some cases of human illness have been reported worldwide (1).

Intoxications which could clearly be attributed to freshwater cyanobacteria have never been published in the Federal Republic of Germany, possibly because surface water usually is not used for drinking water supply or because intoxications are not recognised. Data on the toxicity of cyanobacterial water blooms in the Federal Republic of Germany are lacking. Public interest in the occurrence of toxic cyanobacteria mainly concerns potential health hazards of bathing.

As a first step towards a survey this study investigates the occurrence of microcystin-LR and the species involved in seven Berlin and Brandenburg lakes during summer 1992.

2 MATERIALS AND METHODS

Water samples were taken with a plankton net (mesh size 40 μm) at the shore where the algae were often accumulated by the wind. The algae were further concentrated with a 10 μm mesh size plankton net and freeze-dried.

Microcystin-LR was determined according to Martin et al. (2), modified by Jungman (pers. comm.):

500 mg lyophilised water bloom material was extracted twice with 5 % acetic acid. The crude extract was purified by ODS silica gel chromatography (Bond Elut C-18 cartridges, Analytichem Int.) followed by ion exchange chromatography (ACCELL QMA cartridges, Waters, Eschborn, FRG). HPLC-separation of the toxin was carried out on a Pep-S C2/C18 column (4 x 250 mm, Pharmacia, Freiburg, FRG) at a flow rate of 1 ml min^{-1}. As mobile phase 25 mM CH_3COONH_4 in 65 % methanol/ 35 % H_2O (v/v) was used. Absorbance was monitored from 200 - 300 nm (photodiode array detector 990 Waters, FRG). Corresponding peaks from lake water samples were identified and quantified with authentic microcystin-LR (Calbiochem, FRG).

Phytoplankton biomass was calculated from cell densities counted in an inverted microscope and determination of average cell volumes.

3 RESULTS AND DISCUSSION

Microcystin-LR (retention time 4.5 min) was detected in five lakes in concentrations ranging from 36 to 364 µg/g dry weight (table 1). *Microcystis* spp. dominated in most of the samples, but was often associated with *Planktothrix agardhii* (Wannsee, Templiner See, Schwielow See and Spree) or *Aphanizomenon flos-aquae* (Müggelsee).

Table 1: Content [µg/g dry weight] of microcystin-LR of lyophilised water blooms and dominating cyanobacteria

water body	sampling date	Content [µg/g dry weight] of Microcystin-LR*	cyanobacteria biovolume [cm³/m³]
Wannsee	11.08.92	81	M [5.5 cm³/m³], P [14.9 cm³/m³]
	11.09.92	-	M [68.1 cm³/m³], P [1.1 cm³/m³]
	30.09.92	-	M [16.2 cm³/m³], P [6.5 cm³/m³]
Müggelsee	31.07.92	81	M dominating
	14.08.92	95	M [6.6 cm³/m³], A [1.8 cm³/m³]
	30.08.92	+	M dominating
	08.09.92	71	M [10.6 cm³/m³], A [0.1 cm³/m³]
	25.09.92	55	M [23.1 cm³/m³], A [31.2 cm³/m³]
Templiner See	07.08.92	-	M/ P
	22.09.92	36	M [19.3 cm³/m³], P [3.6 cm³/m³]
Schwielow See	07.08.92	64/toxin ?	M [36.8 cm³/m³], P [6.8 cm³/m³]
	22.09.92	-	M [8.8 cm³/m³], P [11.4 cm³/m³]
Spree (old river branch)	08.09.92	364	M [2.3 cm³/m³], P [0.4 cm³/m³]
Dahme	04.08.92	toxin ?	P dominating
Havel (Aalemannkanal)	30.09.92	toxin ?	M [7.2 cm³/m³], P [0.4 cm³/m³]

*: about 16 % loss during purification process

+: Microcystin-LR, no quantification

cyanobacteria: M: Microcystis spp. (mostly M. aeruginosa), P: Planktothrix agardhii,

         A: Aphanizomenon flos-aquae

toxin ?: substance with different retention time but characteristic microcystin absorption

In Schwielow See another substance with a retention time of 8.4 min and the microcystin-specific absorption maximum at 238 nm was found additionally to microcystin-LR (7.8.92). It is notable that at this time the phytoplankton community of Schwielow See - though dominated by *Microcystis* spp. - also contained a smaller population of *Planktothrix agardhii* (table 1). In the river Dahme (4.8.92) only this substance (retention time 8.4 min) occurred while an almost unialgal water bloom of *Planktothrix agardhii* was present. Therefore it seems likely that this substance was produced by *Planktothrix agardhii*. *Planktothrix agardhii* is known to produce hepatotoxins (3).

In the channel Aalemannkanal of the river Havel (30.9.92) a second unknown substance (retention time 17.1 min) with the same absorption maximum at 238 nm could be detected. At this time *Microcystis* spp. was the dominating species.

Due to their microcystin-characteristic absorption maximum at 238 nm these substances are probably other microcystins. However confirmation of their structural and toxic properties by further analysis is necessary.

These first results indicate that a widespread distribution of hepatotoxic cyanobacteria may be expected in Berlin and Brandenburg lakes. More extensive screening is under way.

REFERENCES

1. G. A: Codd, S. G. Bell and W. P. Brooks, Wat. Sci. Tech., 1989, 21, 1.
2. C. Martin,K. Sivonen, U. Matern, R. Dierstein and J. Weckesser, FEMS Microbiol. Let.,1990,68,1.
3. K. Berg, O. M. Skulberg, R. Skulberg, B. Underdal,and T. Willen, Acta vet. scand., 1986, 27, 440.

Variation of Cyanobacterial Hepatotoxins in Finland

K. Sivonen,[1] M. Namikoshi,[2] R. Luukkainen,[1] M. Färdig,[1]
L. Rouhiainen,[1] W. R. Evans,[3] W. W. Carmichael,[3]
K. L. Rinehart,[2] and S. I. Niemelä[1]

[1]DEPARTMENT OF APPLIED CHEMISTRY AND MICROBIOLOGY, PO BOX 27,
FIN-00014 UNIVERSITY OF HELSINKI, FINLAND

[2]DEPARTMENT OF CHEMISTRY, UNIVERSITY OF ILLINOIS, URBANA, ILLINOIS
61801, USA

[3]DEPARTMENT OF BIOLOGICAL SCIENCES, WRIGHT STATE UNIVERSITY,
DAYTON, OHIO 45435, USA

1 INTRODUCTION

Toxic cyanobacterial (blue-green algal) water blooms commonly occur in eutrophic lakes worldwide[1]. Cyanobacteria produce neurotoxins and peptide hepatotoxins. In a survey conducted in 1985-1987 in Finland 45% of 215 freshwater bloom samples were found to be toxic[2]. Hepatotoxic blooms were more common than neurotoxic blooms as they are more widespread. Cyanobacterial hepatotoxins are either pentapeptides (nodularin) produced by *Nodularia spumigena* in brackish water or heptapeptides (microcystins) found in freshwaters. Microcystins (MCYST) have the general structure *cyclo*(-D-Ala-X-D-MeAsp-Z-Adda-D-Glu-Mdha-), where X and Z are variable L-amino acids, D-MeAsp is D-*erythro*-ß-methylaspartic acid, Mdha is *N*-methyldehydroalanine and Adda is (2S,3S,8S,9S)-3-amino-9-methoxy-2,6,8-trimethyl-10-phenyldeca-4,6-dienoic acid[1]. We have isolated several hepatotoxic strains of the genera *Anabaena*, *Microcystis*, *Oscillatoria* and *Nostoc*[2-11]. The purpose of this study was to determine the structural variation of hepatotoxins in the strains and selected bloom samples. The structural variation and occurrence of these toxins among the cyanobacteria should be known before detection methods are developed.

2 MATERIALS AND METHODS

Cyanobacterial strains were isolated from different lakes in Finland. *Anabaena* strain 83/1[6] and *Microcystis aeruginosa* strain 972[7] originated from Norway and Russia, respectively. Toxin composition of 8 *Anabaena* spp. strains[3-6], 5 *Microcystis* spp. strains and two blooms[7-9], 13 *Oscillatoria agardhii* strains[10], and one *Nostoc* sp. strain[11] was studied. The cells were harvested after 10-12 days of incubation and lyophilized. Toxins were extracted with water or water with organic solvents and purified by high-performance liquid chromatography (HPLC) and thin-layer chromatography (TLC). Amino acid composition was determined by the Waters Pico Tag method and/or gas chromatography (GC) on a chiral capillary column. Structures were assigned by fast atom bombardment mass spectrometry (FABMS), collisionally induced tandem mass spectrometry (FABMS/MS) and ^{1}H NMR[3-11].

3 RESULTS

All strains of *Anabaena*, *Microcystis*, *Oscillatoria* and *Nostoc* studied produced two to ten microcystins simultaneously. From *Anabaena*, *Microcystis*, *Oscillatoria* and *Nostoc*, 27 (19 of which were new), 16 (6 new), 8 (3 new) and 9 (9 new) toxins were found, respectively (Table 1).

Table 1   Microcystins isolated and identified in Finnish cyanobacterial strains.

Genus	No. of strains studied	No. of different toxins	Main toxins
Anabaena	8	27	MCYST-LR, MCYST-RR [D-Asp^3]MCYST-LR, -RR [Dha^7]MCYST-LR, -RR [D-Asp^3,Dha^7]MCYST-LR, -LR or new toxins
Microcystis	5	16	[Dha^7]MCYST-RR [Dha^7]MCYST-LR
Oscillatoria	13	8	[D-Asp^3]MCYST-RR [Dha^7]MCYST-RR
Nostoc	1	9	[ADMAdda^5]MCYST-LR [ADMAdda^5]MCYST-LHar

The structures of 18 different new microcystins were determined and 19 are yet to be identified. MCYST-RR and MCYST-LR and especially their demethylated (amino acid number 3 or 7 or both) variants were the most abundant and frequently occurring toxins among *Anabaena*, *Microcystis* and *Oscillatoria* strains. All *Oscillatoria* strains produced only one major toxin while the strains of *Anabaena* usually produced two to four main toxins simultaneously. *Oscillatoria* strains produced only demethyl microcystins. Microcystins that contain tyrosine were found only in *Microcystis* spp. samples. New variants of microcystins were produced by a *Nostoc* (modified Adda, homoarginine, D-serine in place of D-alanine) and two *Anabaena* strains. *Anabaena* 66 produced four microcystins, three containing homotyrosine and one homophenylalanine. *Anabaena* strain 186 synthesized new compounds, the structures of which have not yet been determined.

4 DISCUSSION

The structures of over 40 cyanobacterial hepatotoxins are known to date - about half of them were characterized in this study. In Finnish freshwaters, cyanobacteria produce a wide variety of microcystins. The variation of toxins as well as the number of species producing these toxins seem to be greater in lakes than in the Baltic Sea. Only nodularin, a pentapeptide hepatotoxin, produced by *Nodularia spumigena* has been found in the Baltic Sea to date[12]. In Japan, 35 *Microcystis aeruginosa* and *M. viridis* strains were studied and only three microcystins (MCYST-LR, -RR and -YR) were found[13]. In Finland the demethyl microcystins were especially common. Structurally similar toxins are produced by different species but some strains produce only new varieties of microcystins. The qualitative and quantitative variations of microcystins were greatest among *Anabaena* and lowest among the isolates of *Oscillatoria*. This study indicates that in some cases the use of HPLC as a method to detect and quatitate toxin content of natural blooms will prove difficult, since blooms may consist of several toxic species/strains[2] and several different toxins might be produced by each strain. The availability of a full range of toxin standards for comparison and identification would be necessary when HPLC is used to detect freshwater microcystins. Studies of biogenesis and genetics of cyanobacterial microcystins will be needed to reveal why so many varieties of these compounds are produced at a time and why certain compounds are found as the major toxins.

REFERENCES

1.   W.W. Carmichael, J. Appl. Bact., 1992, 72, 445.
2.   K. Sivonen, S.I. Niemelä, R.M. Niemi, L. Lepistö, T.H. Luoma and L.A. Räsänen. Hydrobiologia, 1990, 190, 267.
3.   K. Sivonen, M. Namikoshi, W.R. Evans, W.W. Carmichael, F. Sun, L. Rouhiainen, R. Luukkainen and K.L. Rinehart, Appl. Environ. Microbiol., 1992, 58, 2495.
4.   M. Namikoshi, K. Sivonen, W.R. Evans, W.W. Carmichael, L. Rouhiainen, R. Luukkainen and K.L. Rinehart, Chem. Res. Toxicol., 1992, 5, 661.
5.   M. Namikoshi, K. Sivonen, W.R. Evans, W.W. Carmichael, F. Sun, L. Rouhiainen, R. Luukkainen and K.L. Rinehart, Toxicon, 1992, 30, 1457.
6.   K. Sivonen, O.M. Skulberg, M. Namikoshi, W.R. Evans, W.W. Carmichael and K.L. Rinehart, Toxicon, 1992, 30, 1465.
7.   K. Sivonen, M. Namikoshi, W.R. Evans, B.V. Gromov, W.W. Carmichael and K.L. Rinehart, Toxicon, 1992, 30, 1481.
8.   J. Kiviranta, M. Namikoshi, K. Sivonen, W.R. Evans, W.W. Carmichael and K.L. Rinehart, Toxicon, 1992, 30, 1093.
9.   R. Luukkainen, M. Namikoshi, K. Sivonen, K.L. Rinehart and S.I. Niemelä, Toxicon, 1994, 32, 133.
10. R. Luukkainen, K. Sivonen, M. Namikoshi, M. Färdig, K.L. Rinehart and S.I. Niemelä, Appl. Environ. Microbiol., 1993, 59, 2204.
11. K. Sivonen, M. Namikoshi, W.R. Evans, M. Färdig, W.W. Carmichael and K.L. Rinehart, Chem. Res. Toxicol., 1992, 5, 464.
12. K. Sivonen, K. Kononen, W.W. Carmichael, A.M. Dahlem, K.L. Rinehart, J. Kiviranta and S.I. Niemelä, Appl. Environ. Microbiol., 1989, 55, 1990.
13. M.F. Watanabe, S. Oishi, K.-I. Harada, K. Matsuura, H. Kawai and M. Suzuki, Toxicon, 1988, 26, 1017.

A Method for the Detection of Cyanobacterial Peptide Toxins by HPLC

J. Rositano and B. C. Nicholson

AUSTRALIAN CENTRE FOR WATER QUALITY RESEARCH, PRIVATE MAIL BAG, SALISBURY, SOUTH AUSTRALIA 5108

1 INTRODUCTION

Recent awareness of the significance of toxic cyanobacteria to water supply has led to increased research and monitoring programs for toxic cyanobacteria in our laboratories. Central to this has been method development for the analysis of cyanobacterial peptide toxins in water and scum samples.

The analytical procedure described is dependent upon the nature of the sample (Figure 1). Toxins in samples containing no live cells are concentrated by solid-phase C18 extraction followed by HPLC analysis. Samples containing less than 1×10^6 cells/mL are sonicated to lyse cells and analysed as for water samples. Scum samples having greater than 1×10^6 cells/mL are freeze-dried, the toxin extracted and analysed directly by HPLC.

Materials and Methods

HPLC Analyses were performed with a Waters Associates liquid chromatograph equipped with a model 501 pump, 991 photodiode array detector set at 240nm and a 917 auto-sampler. The column was a Brownlee ODS 5 μm analytical column, 220 mm x 4.7 mm ID. The mobile phase was an isocratic solution of 27% acetonitrile/pH 7, 0.1M potassium dihydrogen orthophosphate/sodium hydroxide buffer run at 1.0 mL/min[1].

Cyanobacterial Materials. Freeze-dried *Microcystis aeruginosa* from Mt. Bold Reservoir, South Australia (10.2.88) and fresh cellular material from Centennial Park Lake, New South Wales (24.5.93) were used as the source of toxic cyano-bacteria. The material from Mt. Bold contained microcystin-LR and -LA[1] whilst the material from Centennial Park contained only one peak with a UV absorption maxima at 240 nm which co-eluted with microcystin-LR. Nodularin and Microcystin-LR standard were purchased from Calbiochem Corp., California.

Solid Phase Extraction. A Waters 500 mg C18 solid phase extraction cartridge attached to a vacuum manifold (Alltech) was primed with methanol and water (15 mL of each). The sample was vacuum filtered through the cartridge at approx. 10 mL/min. The cartridge was washed with water, 10% methanol, 20% methanol (10 mL of each) and the toxin eluted with 100% methanol (10mL).

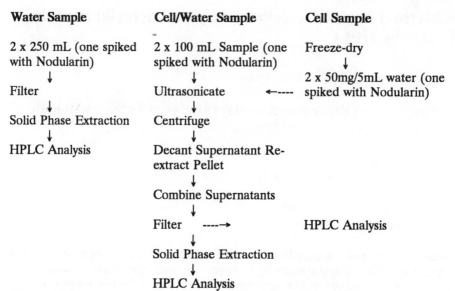

Figure 1. Flow diagram of extraction procedures for Water, Cell/Water, and Cell sample analysis.

The methanol extract was evaporated to dryness at 40°C under nitrogen and the dry extract was reconstituted to 1 mL with 50% methanol/water.

Cell Extraction. Samples of fresh intact cells of 100 mL volume were sonicated for 5 minutes using a Branson sonifier 250 and centrifuged at 4000 rpm for 20 minutes. The extraction was repeated in 20 mL distilled water and the supernatants combined. Freeze-dried cells in distilled water (50 mg/5 mL) were sonicated for 2 minutes using a Branson sonifier with microtip attachment. The lysed cells were centrifuged (4000 rpm, 20 minutes) and re-extracted in 5 mL of distilled water. Sample extracts were filtered through 0.45 μm membranes before solid phase extraction or HPLC analysis.

2   RESULTS AND DISCUSSION

To determine the precision of recovery for the analysis of water samples, untreated water from Myponga Reservoir South Australia, (DOC 9.8 mg/L, pH 7.3, TDS 270 mg/L) and distilled water were spiked with various quantities of microcystin-LR and analysed six times. Results, given in Table 1 indicate that reproducible results are obtained for concentrations ranging between 1-40 μg/L in distilled water. Reproducibility is poorer at low toxin concentrations in water of high DOC due to the naturally occurring organic matter concentrated by the extraction cartridge which interferes with HPLC analysis.

Water from Centennial Park Lake containing 2×10^6 cells/mL and dilutions of this material was extracted according to the Cell/Water sample procedure (Figure 1) in order to determine linearity of extraction over the range tested. Each extraction was repeated five times and spiked with 2 μg nodularin as the internal standard. Linearity over the range 1×10^5 to 2×10^6 cells/mL was achieved (Figure 2).

Table 1. Recovery of Microcystin-LR from Water

Microcystin-LR Conc. (μg/L)	% Recovery Reservoir Water	Distilled Water
40	98 (2.5)	89 (3.7)
20	88 (3.5)	100 (7.7)
10	93 (3.2)	91 (10.0)
5	90 (6.6)	99 (1.6)
2	88 (10.5)	100 (7.1) 88*(8.0)
1	122 (21)	79 (11.5)

() = standard deviation, * = Nodularin recovery.

Freeze-dried *M. aeruginosa* from Mt. Bold Reservoir was extracted according to the method for Cell sample analysis outlined in Figure 1. Five replicate extractions gave 153 μg total toxin content, having a recovery of 81% (based on the recovery of nodularin) and standard deviation of 10 μg (6.0%). Toxin concentrations in all experiments were determined by the comparison of peak areas with those of standards where available. Where toxin standards were not available concentrations were expressed as microcystin-LR equivalents.

3   SUMMARY

The results obtained in all cases are highly satisfactory for the extraction procedures outlined. Toxin recovery as determined from the internal standard was above 80% for samples having a concentration above 1 μg/L and efficiency of toxin extraction was independent of cell numbers in the range 1 x 10^5 to 2 x 10^6 cells/mL.

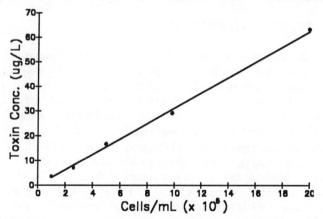

Figure 2. Cell/Water extraction efficiency as a function of cell numbers.

REFERENCES

1.   D.J. Flett and B.C. Nicholson, "Toxic Cyanobacteria in Water Supplies." Report No. 26, Urban Water Research Association of Australia, Melbourne (1991).

Release and Degradation of Microcystin during a *Microcystis Aeruginosa* Bloom in a Freshwater Reservoir

C. S. Dow, U. K. Swoboda, and P. Firth

DEPARTMENT OF BIOLOGICAL SCIENCES, UNIVERSITY OF WARWICK, COVENTRY
CV4 7AL, UK

1   INTRODUCTION

Several naturally occurring cyanobacteria produce toxins which are lethal to livestock, wildlife and pets[1-4]. The isolation, purification and characterisation of toxins produced by these cyanobacteria is particularly important both in terms of the presence of the toxins in the algal blooms and the possible release of these toxins into water especially with the increasing frequency of occurrence of blooms in reservoirs serving both recreational purposes and as sources of potable water. The occurrence of cyanobacterial toxins in potable water may present a serious health hazard to humans if very low levels of these toxins are consumed over a long period of time, contributing to chronic liver problems such as necrosis or liver tumours[5].

In this study we report the isolation, purification and characterisation of a toxin produced by a bloom of *Microcystis aeruginosa* from Cropston reservoir in the English Midlands and follow the fate of toxin in reservoir water over a period of time.

2   RESULTS AND DISCUSSION

Figure 1 shows the high performance liquid chromatogram of *Microcystis aeruginosa* cell extract from Cropston reservoir. The lethal dose was 1 mg dry weight cells per 20g mouse. The fraction eluting at t_R25.349 minutes was spectrally pure, toxic by mouse bioassay and had an excellent match to microcystin-LR (similarity of 0.99999 and a dissimilarity of 0.00472 to microcystin-LR using the Varian software, "Polyview"). 25 μg of this eluate fraction was sufficient to kill a 20g mouse within 2 hours. MALDI mass spectrometry showed this peak to have a relative molecular mass of 998 which was identical to the molecular mass of microcystin-LR recovered from water into which it had been spiked. The raw water sample collected at the same time as the toxic cyanobacterial biomass was found to be non-toxic although a concentrate, equivalent to 10 litre of water (concentrated by passage through a Sep- Pak C_{18} cartridge), had been injected into each mouse. Moreover, no toxic peak fraction was identified in this sample.

In order to evaluate the possibility of free toxins being released into the water by the decaying cyanobacterial cells, the stability of toxins released into the water and the effect of high background caused by interfering components of the reservoir water on the isolation and identification of hepatotoxins, a large volume Cropston water containing a high number of toxic *Microcystis aeruginosa* cells (as per the biomass analysed in Figure 1) was analysed over a period of 61 days. During this time the cyanobacterial biomass degraded yielding a high concentration of cell debris. These samples were subsequently processed and the cell-free water analysed for the presence of microcystins. With the high concentration of decaying toxic cells it was expected to detect "free" hepatotoxins in the water by high performance liquid chromatography of water concentrates assayed on days 18 and 54 respectively (although there was a very high background, which increased with time, this did not adversely affect the detection of spiked toxin standards). However, no peaks were resolved with a good match to the toxin standards.

Figure 1.   HPLC elution profile of a cell lysate of *Microcystis aeruginosa* with the toxic fraction eluting at t_R25.349 minutes. The spectral overlay of this peak gave an excellent match to microcystin-LR. Chart speed 0.52 cm min^{-1}; attenuation 141; zero offset 3%. No toxic peaks were detected by HPLC analysis of 10 litre raw water concentrates of the same sample.
Chart speed 0.52 cm min^{-1}; attenuation 128; zero offset 5%.

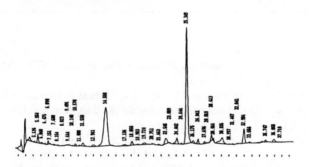

To validate the analysis and to confirm the sensitivity and specificity of detection, 61 day old raw water from Cropston was spiked with microcystin-LR and nodularin at concentrations ranging from 0.2 to 10µg litre^{-1}. Despite the high background, the toxins were readily detected by HPLC analysis. Concentrations as low as 200 ng toxin per litre of water were detected. MALDI mass spectral analysis of microcystin-LR spiked into, and recovered from, Cropston water had a relative molecular mass of 998 i.e. confirmation of microcystin-LR (Figure 2). In order to monitor the raw water in the Cropston reservoir itself, water was collected at regular intervals and assayed for the presence of hepatotoxins. No "free" toxins were detected. The question of the stability of the toxins remains to be addressed but current data indicates that the hepatotoxins remain associated with the cell debris, are readily degraded by the indigenous microbial population and /or removed by dissolved organic carbon and particulate material.

Figure 2.   MALDI mass spectrum of microcystin-LR recovered after spiking of Cropston reservoir water. The peak at 998 is microcystin-LR, those at 1020, 1041 and 1058 are protonated and/or metal ion (Na^+, K^+ or Li^+) adducts of the toxin.

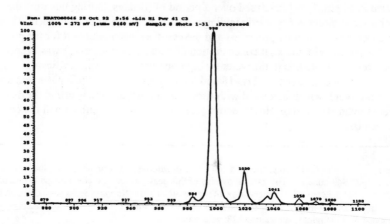

REFERENCES

1.   W.W. Carmichael, "Handbook of Natural Toxins", ed. A.T. Tu, Marcel Dekker, New York, 1988, Vol. 3, p121.

2.   C.S. Dow, U.K. Swoboda and V. Howells, "Recent Advances in Toxinology Research", ed. C.Gopalakrishnakone and C.K. Tan, National University of Singapore, 1992, Vol. 3, p323.

3.   I.R. Falconer, "Environmental Toxicology and Water Quality", 1991, 6, p511.

4.   K.I. Harada, K. Ogawa, Y. Kimura, H. Murata, M. Suzuki, P. Thorn, W.R. Evans and W.W. Carmichael, "Chemical Research and Toxicology", 1991, 4, p534.

5.   U.K. Swoboda, C.S. Dow and A. Wilson, "Recent Advances in Toxinology Research", ed. C.Gopalakrishnakone and C.K. Tan, National University of Singapore, 1992, Vol. 3, p307.

ACKNOWLEDGEMENT

This research was sponsored by Severn Trent Water, Birmingham, UK.
Their help and involvement, especially that of Les Markham and Helen Picket, is gratefully acknowledged.

Characterization of Hepatotoxins from Freshwater *Oscillatoria* Species: Variation in Toxicity and Temporal Expression

J. Chaivimol, U. K. Swoboda, and C. S. Dow

DEPARTMENT OF BIOLOGICAL SCIENCES, UNIVERSITY OF WARWICK, COVENTRY
CV4 7AL, UK

1 INTRODUCTION

Heavy blooms of cyanobacteria appear widely in nutrient-rich fresh and brackish water. Several of these bloom forming species are known to produce toxins which are responsible for the death of livestock and wildlife in many parts of the world.[1-3] These toxins include peptide hepatotoxins and alkaloid neurotoxins.

It is now known that the toxicity of a particular cyanobacterial species varies between sites (and within a particular site), on a seasonal, weekly and even daily basis [4,5] and although the growth of cyanobacteria is influenced by environmental factors such as light intensity, temperature and pH, the environmental and physiological conditions required for the induction of toxin expression are still unclear.

Lower Shustoke reservoir, a freshwater reservoir in the English Midlands was monitored regularly during March to November, on a bi-weekly basis. This study reports the occurrence and toxicity of hepatotoxins produced by an *Oscillatoria* sp. which was the predominant species found in this reservoir. The environmental samples were analysed by HPLC and, for toxicity, by mouse bioassays. Toxicity of the cyanobacterial biomass has been shown to be variable in both qualitative and quantitative terms, presumably in response to cellular and/or environmental stimuli. The molecular basis of the structural variations of the toxins has yet to be elucidated.

2 RESULTS AND DISCUSSION

The predominant cyanobacterial species in Lower Shustoke reservoir during a 2 year monitoring period was always an *Oscillatoria* sp. Throughout this time, the biomass, as determined by mouse bioassay, was always toxic and the symptoms were characteristic of hepatotoxins. However, the concentration of cell biomass required to cause death of mice varied over the sampling period, with death resulting from within a few minutes to several hours, after intraperitoneal injection. HPLC analysis of lyophilised cell biomass, following extraction with 5% acetic acid and purification via a Sep-Pak C18 cartridge, using a reversed phase C18 column and a linear gradient of 30 – 60 % acetonitrile containing 0.05% trifluoroacetic acid, indicated 3 toxic peaks which had spectral similarity close to nodularin and microcystin(s) (Figure 1).

Figure 1   HPLC eluate profile of cyanobacterial biomass from Lower Shustoke (2/9/92). The peaks, t_R 16.763, 17.417 and 18.097 were all toxic in mouse bioassays. Chart speed 0.51 cm min^{-1}, attenuation 25, zero offset 0%.

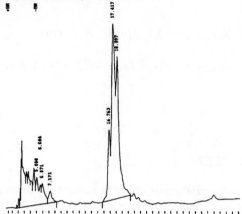

The relative molecular masses determined by matrix assisted laser desorption ionisation mass spectrometry were 1048, 1027 and 984 respectively (Figure 2).

Figure 2   MALDI mass spectrum of eluate peak t_R 16.76 from Lower Shustoke which had good spectral similarity to nodularin.

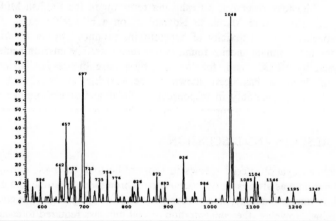

Each peptide had a distinct toxicity level in mouse bioassays. Moreover, there was a temporal variation in the concentration of these toxins, on a per cell basis, during the monitoring period (Figure 3). The temporal variation in the toxicity of the cells was dependant upon the ratio of the toxic peptide components present in the cells. This variability may be dictated by intrinsic and/or environmental factors such as gene expression, the physiological state of the cell, nutrient concentration, water temperature and/or light intensity.

Figure3 Temporal expression of cyanotoxins in an *Oscillatoria* sp.

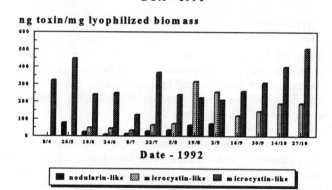

ng toxin/mg lyophilized biomass

Date - 1991

ng toxin/mg lyophilized biomass

Date - 1992

nodularin-like ▨ microcystin-like ▨ microcystin-like

Despite the highly toxic nature of the biomass and the detection limit of the hepatotoxins being of the order of 170/200 ng per litre of raw water (resolution per 10 µl injection of 40 ng), "free" toxins could not be detected in any raw water samples. The non-toxicity of these water samples was confirmed by mouse bioassay.

REFERENCES
1   D.P. Botes, "Mycotoxins and Phycotoxins, Bioactive Molecules", Elsevier, Amsterdam, 1986, Vol. 1, p161.
2   W.W. Carmichael, "Handbook of Natural Toxins", ed. A.T. Tu, Marcel Dekker, New York, 1988, Vol. 3, p12.
3   P.R. Gorham and W.W. Carmichael, Pure Appl. Chem., 1979, 52, p165.
4   C.S. Dow, U.K. Swoboda and V. Howells,"Recent Advances in Toxinology Research", ed. C. Gopalakrishnakone and C.K. Tan, National University of Singapore, 1992, Vol 3, p323.
5   U.K Swoboda, C.S. Dow and A. Wilson,"Recent Advances in Toxinology Research", ed. C. Gopalakrishnakone and C.K. Tan, National University of Singapore, 1992, Vol 3, p307.

ACKNOWLEDGEMENT
This research was sponsored by Severn Trent Water, Birmingham, UK.

Expression of Cyanotoxins in Environmental Biomass Containing Species of *Oscillatoria*

U. K. Swoboda and C. S. Dow

DEPARTMENT OF BIOLOGICAL SCIENCES, UNIVERSITY OF WARWICK, COVENTRY CV4 7AL, UK

1 INTRODUCTION

Toxic blooms are known to be produced by the filamentous cyanobacterium *Oscillatoria* and the coccoid *Microcystis*[1]. To date the majority of these toxins have been identified as microcystin-like hepatotoxins. Over fifty such cyclic peptides have been isolated from cyanobacteria including other species in the genera *Anabaena, Nodularia* and *Nostoc* .

Recently it has been reported that two different strains of *Oscillatoria*, in addition to *Aphanizomenon* and *Anabaena*, produce neurotoxins which have been identified as anatoxin-a and a methylene homologue of anatoxin-a termed homoanatoxin[2-3]. Heavy blooms of cyanobacteria in freshwater reservoirs in the English Midlands were regularly monitored during March to November during 1991 and 1992, on a bi-weekly basis. Biomass from many of these has been found to be toxic[4] and the toxicity has been shown to vary in both qualitative and quantitative terms.

We report here the variation in toxins produced by *Oscillatoria* sps. from four Midland reservoirs; variation in the temporal expression of resolvable peptide toxins produced by the *Oscillatoria* sp. from Lower Shustoke reservoir and investigation of the marine species *Oscillatoria erythrea* for the presence of peptide hepatotoxins.

2 RESULTS AND DISCUSSION

The predominant cyanobacterial species in Lower Shustoke was an *Oscillatoria* sp.. Throughout the period of evaluation, the cell biomass from this reservoir was always toxic in mouse bioassays and gave symptoms characteristic of hepatotoxins. However, the toxicity of the cells varied over the sampling period, with death resulting from within a few minutes to several hours. There was a temporal variation in the expression of the three toxic peaks with molecular masses of 1048, 1027 and 984. Details of this study are disscussed in the paper by Chaivimol *et al.*[7] Earlswood Lake also supported the growth of a toxic *Oscillatoria* sp. which showed 3 major peaks following HPLC (Figure 1) but with molecular masses of 1023, 974 and 913 respectively. However, only peaks 1 and 3 were toxic in mouse bioassays.

Figure 1. HPLC elution profile of cell extract from the *Oscillatoria* sp. from Earlswood Lake. Eluate peaks t_R 16.392 and 18.302 were toxic. Eluate peak t_R 16.883 was non-toxic. Chart speed 0.51 cm min^{-1}; attenuation 160 ; zero offset 0%

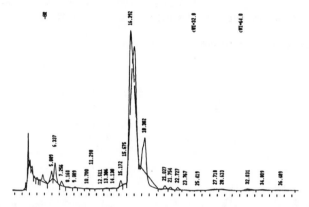

The predominant species during a cyanobacterial bloom in Linacre reservoir was another *Oscillatoria* sp. Its toxicity was dose-related, with cell lysates as low as 1mg dry weight of cells being sufficient to kill 20 g mice within 3 hours post injection. The toxicity of the cells was due to a major peak eluting at 21.553 minutes during HPLC analysis and was spectrally very similar to microcystin-LR.

The *Oscillatoria* sp. collected during a bloom in Thornton reservoir in 1991, was neurotoxic and manifested symptoms such as reduced activity, ataxia, piloerection, diarrhoea, paralysis of hind limbs/reduced reflexes and convulsion when injected into mice intraperitoneally. The severity of symptoms was dose-dependent. However, death did not occur until after approximately 24 hours even with 13.5 mg dry weight cell lysate being administered per 20 g mouse (equivalent to 350 mg dry wt kg^{-1} body weight) although the symptoms were manifested shortly after injection. From spectral data it was evident that no peak fractions had similarity to anatoxin-a or to the hepatotoxin standards (Figure 2a and b). The major single peak with a t_R of 37.85 minutes was non-toxic when injected into mice at a dose of 4 µg per 20 g mouse. The toxicity of the sample was caused by a compound(s) which was not identified by our assay system.

Intraperitioneal injections into mice of *Oscillatoria erythrea*, a red marine cyanobacterium collected during a bloom off Heron Island, Great Barrier Reef, Australia, manifested symptoms identical to those reported for ciguatoxin. The neurological symptoms included reduced activity, paralysis of limbs, piloerection, unusual gait, diarrhoea and breathing difficulties in mice. The severity of the symptoms was dose-dependent but even as much as 30 mg lyophilised cell lysates were insufficient to kill 20 g mice and all the animals tested subsequently recovered. These cells did not exhibit the same severity of symptoms as those

Figure 2. HPLC chromatograms of (a) anatoxin-a .HCl, microcystin-LR and nodularin; (b) *Oscillatoria* sp. from Thornton reservoir.
The toxins were eluted in a linear gradient of 3-60% acetonitrile containing 0.05% trifluoroacetic acid at a flow rate of 1ml minute[-1]. Anatoxin-a.HCl eluted at 14.873 minutes, while the hepatotoxins nodularin and microcystin-LR had retention times of 39.737 and 42.788 minutes respectively.
Chart speed 0.35 cm min[-1]; attenuation 30; zero offset 18%.

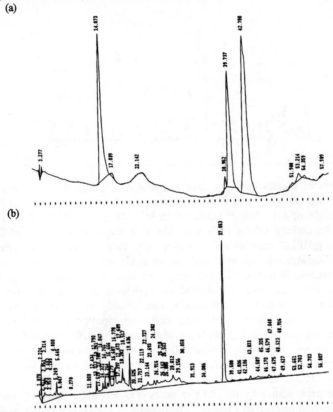

reported for samples collected in 1988 and 1989 where death ensued[5] but were similar to those collected in 1987 (Hahn, S.T., personal communication). Neither anatoxin-a nor hepatotoxins were identified by HPLC analysis. All the major eluate peaks were collected and shown to be non-toxic by mouse bioassay. It has now been shown by Hahn and Capra (1992) that the toxicity of *O. erythrea* is due to a neurotoxic polycyclic ether which is chemically indistinguishable from ciguatoxin and the species probably acts as a potential elaborator of a ciguatoxin-like compound in the tropical marine biota.

The variability of cyanotoxins produced by the different *Oscillatoria* strains is not surprising since over fifty microcystin-like hepatotoxins alone have been reported to date with several of these having been isolated from the same cyanobacterial species[1]. It is apparent that toxin expression varies over the year and within strains which produce more than one toxin. The variability in the toxicity of these cells is dependent upon the ratio of the component peptides since each peptide has a distinct toxicity level in mouse bioassays. This variability in the toxicity of the different toxins is due to small structural changes, as indicated by the small changes in their relative molecular masses.

REFERENCES
1.   W.W.Carmichael, J. Appl. Bact., 1992, 72, p445.
2.   K.Sivonen, K.Himberg, R.Lukkainen, S.Niemela, G.K. Poon and G.A. Codd, Toxicity Assessment, 1992, 4, p339.
3.   O.M.Skulberg, W.W.Carmichael, R.A. Anderson, S. Matsunaga, R.L.Moore, and R. Skulberg, Environmental Toxicology and Chemistry,1992, 11, p321.
4.   Dow, C.S., Swoboda, U.K. and Howells, V.,"Recent Advances in Toxinology Research", ed. C. Gopalakrishnakone and C.K. Tan, National University of Singapore, 1992, Vol.3, p323.
5.   S.T.Hahn and M.F.Capra, Food Additives and Contaminants,1992, 9, p351.
6.   U.K.Swoboda, C.S.Dow, and A.Wilson, "Recent Advances in Toxinology Research", ed. C. Gopalakrishnakone and C.K. Tan, National University of Singapore, 1992, Vol.3, p307.
7.   J. Chaivimol, U.K. Swoboda, and C.S. Dow, this volume, p161-163.

ACKNOWLEDGEMENT
This research was sponsored by Severn Trent Water, Birmingham, UK.
We would like to thank Les Markham and Helen Picket for providing us with environmental samples and greatly appreciate the skilful technical assistance of Peter Firth and Nicola Smith.

A Simple and Rapid Method for Extraction of Toxic Peptides from Cyanobacteria

Nina Gjølme and Hans Utkilen

NATIONAL INSTITUTE OF PUBLIC HEALTH, DEPARTMENT OF ENVIRONMENTAL MEDICINE, GEITMYRSVEIEN 75, N-0462 OSLO, NORWAY

1 INTRODUCTION

Several methods (1, 2, 3, 4 and 5) have been described for extraction, concentration and quantification of the toxic peptides from Microcystis and other cyanobacteria. All these methods are variations of a basic procedure which includes freeze drying of a liquid sample, extraction lasting from 5-120 min, centrifugation and concentration on C_{18} cartridges prior to quantification on HPLC. These methods are time consuming, and water samples have to be sent to a laboratory.

A simple and rapid method for concentration of cell material and extraction of peptide toxins is presented, a method that is useful both for laboratory and field sampling, and which eliminates transport of liquid samples from the field to a laboratory.

2 MATERIALS AND METHODS

The organism used to develop the method was Microcystis aeruginosa (CYA 228/1) obtained from the Norwegian Institute for Water Research, Oslo.

The cell material was collected on Whatman GF/C (47 mm) filters by suction, and the filters were frozen. The filtrates were collected for control of toxin content by HPLC, because some toxin could have leached from dead and decayed cells into the water or culture medium during cell growth. After thawing the cell-covered filters, the peptide toxins were extracted by passing 5% acetic acid or water through the filters by suction, and the extracts were analyzed on HPLC. The HPLC used for analysing filtrates and extracts was fitted with an internal surface reverse phase (ISRP) column (Pinkerton, USA). The mobile phase was 0.1 M KH_2PO_4 with added acetonitrile (12%), pH 6.8 and flow 1 ml/min. The detector was set at 238 nm. Standards for toxin quantification were isolated and purified from Microcystis aeruginosa CYA 228/1.

As a control, liquid samples equivalent to the

filtered ones were frozen, thawed and acetic acid added to
5%, before a 90 min extraction with stirring at 4°C. The
extracts were centrifuged and the supernatants were applied on
C_{18} (Sep-Pac, Waters) cartridges and the toxins were eluted
with methanol. The methanol eluates were analyzed on HPLC.

3 RESULTS AND DISCUSSION

Samples of a <u>Microcystis</u> culture were filtered on GF/C
filters. The cell-covered filters were placed in Petri dishes
and treated in different ways. Some filters were frozen (-20
°C) immediately after filtering, while they still were wet,
others had 0.5-5 ml of water added and were frozen. One group of
filters was dried overnight at room temperature, 0-5 ml water
was added the next day and the filters were frozen, while some
where dried overnight and extracted without freezing. After
thawing the frozen filters, toxin extraction was performed by
sucking a known volume of 5% acetic acid through the filters.
When the Petri dish contained water after thawing, this was
also passed through the filter and the dish was washed with
some of the 5% acetic acid. The toxin content in the extracts
was determined on HPLC. The results revealed that freezing a
wet filter without addition of water was sufficient for
extraction of the toxins. For dried filters, adding 0.5 ml
water before freezing was found to be sufficient for recovery
of the cellular toxins. The remaining extracts were applied on
C_{18} cartridges, eluted with methanol and the eluates were
analyzed on HPLC. After C_{18} treatment the toxin recovery in
the filter extracts was found to be identical to the controls.

A comparison of the toxin contents in the 5% acetic
acid extracts before and after a C_{18} treatment revealed that
about 16% of the toxin was lost during the C_{18} treatment.

An examination of the necessary extraction volume
and time was performed by filtering 200 ml of a <u>Microcystis</u>
culture. With the cell density of the culture used 200 ml was
about the maximal culture volume that could be filtered. After
freezing and thawing the filters, 5-25 ml of 5% acetic acid
was sucked slowly (20 sec.) or rapidly (few sec.) through the
filters to extract the toxins. All the filters were
reextracted with 10 ml. Table 1 shows that passing 10 ml
slowly through the filter resulted in the best recovery of the
toxin. Since no toxin was found in any of the second extracts,
a slow extraction with 10 ml 5% acetic acid is sufficient for
toxin extraction.

An important advantage of the filter method is that
the cellular toxins are concentrated directly, since large
sample volumes can be filtered and extracted with small
volumes.

Toxin extraction with distilled water was also
examined. The results revealed that the water extracts
contained less contamination, and a better peak separation
was obtained than for the 5% acetic acid extracts, while the
same toxin values were obtained by both extractions. At the

present conference it was said that the extraction of the
peptide toxins is not complete with water or 5% acetic acid
and that a methanol extraction should be used. We are
therefore in the process of examining the effect of various
methanol concentrations on the extraction of toxin from cells
on the GF/C filters. The results will be published later.

Table 1. Extraction of peptide toxin from <u>Microcystis</u>
<u>aeruginosa</u> as a function of extraction time and volume.

Extraction volume, ml	Extracted toxin, μg/ml culture after	
	3-5 sec.	20 sec.
5	89 ± 4	96 ± 9
10	85 ± 7	114 ± 2
15	--	109 ± 3
20	93 ± 9	--
25	92 ± 5	104 ± 6

Toxin yields are given as the mean of 3 determinations ±
standard deviation.

The method has also been applied to <u>Oscillatoria</u> and <u>Anabaena</u>,
containing peptide toxins, with the same results.

4 CONCLUSIONS

Based on our findings, we propose the following procedure for
sampling and extraction of toxic peptides from cyanobacteria:

Samples are filtered on GF/C filters, the wet
filters are placed in Petri dishes and frozen. Field samples
are filtered, dried and sent together with some of the
filtrates to a laboratory. Dried filters are rehydrated with
0.5 ml water and frozen. After thawing, 10 ml water or methanol/water
(depending on further findings) is passed slowly (20 sec.)
through the filters for toxin extraction. The extracts and
filtrates are analyzed on HPLC.

The method allows samples for toxin determination to
be mailed on filters to a laboratory, since no loss of toxin
was found after storing samples on filters at room temperature
in the dark for some weeks. The method also results in a
direct and rapid concentration of cellular toxin in a water
sample.

5 REFERENCES

1. W.P. Brooks and G.A. Codd, Letters Appl. Microbiol. 1986 <u>2</u> 1-3.
2. K.J. Harada, K. Matsuura, M. Suzuki, H. Oka, M.F. Watanabe, S. Oishi, A.M. Dahlem, V.R. Beasley and W.W. Carmichael, J. Chromatography.1988 <u>448</u> 275-283.
3. C. Martin, K. Sivonen, U. Matern, R. Dierstein and J. Weckesser, FEMS Microbial. Lett. 1990 <u>68</u> 1-6.
4. J.A.O. Meriluoto and J.E. Eriksson. J. Chromatography 1988 <u>171</u> 867-874.
5. Ø.Østensvik, O.M. Skulberg and N.E. Søli. The water environment: Algal toxins and health. Ed. W.W. Carmichael, Plenum Press, New York, 1982. pp. 315-324.

Phosphatase Assay as a Determinant of Hepatotoxin Toxicity

J. Chaivimol, U. K. Swoboda, and C. S. Dow

DEPARTMENT OF BIOLOGICAL SCIENCES, UNIVERSITY OF WARWICK, COVENTRY
CV4 7AL, UK

1   INTRODUCTION

Microcystis aeruginosa, the most common toxic cyanobacterium so far studied, produces hepatotoxins which are a related family of cyclic heptapeptides with molecular weights ranging from 980 to 1035.[1]   They cause poisoning of livestock, wildlife and have also been implicated with human illness associated with the proliferation of cyanobacteria in sources of potable water.[2]   Recently, it has been shown that microcystin produced by *Microcystis aeruginosa* is a potent and specific inhibitor of the protein phosphatases 1 and 2A.[3-4] We report here the effect on protein phosphatases of (i) pure microcystin-LR, microcystin-RR and nodularin, (ii) whole cyanobacterial cell extracts, (iii) toxins extracted from environmental cell biomass and (iv) raw water samples.

2   RESULTS AND DISCUSSION

Phosphorylase phosphatase activity was strongly inhibited by microcystin-LR, -RR and nodularin when assayed against phosphorylase *a* as substrate and liver homogenate of Balb C mice as the source of phosphatase enzymes. The IC_{50} (the concentration causing 50% of the maximum inhibition of phosphatase activity) of microcystin-LR, -RR and nodularin were 2.4 nM, 2.8 nM and 2.3 nM respectively, whereas the IC_{50} of okadaic acid, a potent protein phosphatase inhibitor and potent tumour promotor, was 24.5 nM. Although microcystin-LR was 10 times less toxic than microcystin-RR when injected intraperitoneally into mice (LD_{50} of microcystin-LR is 3μg and the LD_{50} of microcystin-RR is 30μg), they both have approximately the same IC_{50}. This difference may be due to either permeability effects, modification of microcystin-RR during uptake in animal systems and/or differential activation of the two microcystins *in vivo*.

Figure 1 shows the inhibitory effects of several environmental samples. There was a perfect correlation in toxicity results between mouse bioassays, HPLC analysis and phosphatase assays of the hepatotoxin standards, whole cells and toxic peak fractions collected following HPLC analysis of natural blooms from freshwater reservoirs.

Figure 1 Inhibitory effect of the whole cell biomass of an *Oscillatoria* sp. and a *Microcystis* sp. on the activity of protein phosphatase.

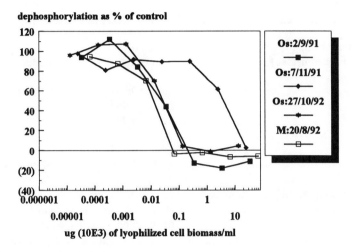

The IC_{50} of the whole cells were in the range of 12.9 to 26.8 mg of lyophilised cells ml^{-1} (Table 1) and those of toxic fractions separated by HPLC were in the range of 3.9 to 31.1 nM (Table 2). Raw water concentrates equivalent to 200 ml of "cell free" filtrate did not have any effect on the activity of the phosphorylase phosphatases.

Table 1 : Effects of peptide toxins from environmental biomass

Reservoir	Date	Species	IC_{50} (nM)
Lower Shustoke	2/09/91	*Oscillatoria*	24.5
	7/11/91	*Oscillatoria*	>3500
	27/10/92	*Oscillatoria*	26.8
Cropston	20/8/92	*Microcystis*	12.9

IC_{50} (nM) : inhibition of phosphatase activity, μg of lyophilized cells ml^{-1}

Table 2 :   Effects of defined HPLC eluate fractions on protein phosphatase activity

Cyanobacterial sp. (reservoir)	Date	Spectral match	IC_{50} (nM)
Oscillatoria (Lower Shustoke)	8/4/92	mLR : 0.998	3.9
		no match	no effect
	8/7/92	no match	no effect
		mRR : 0.993	8.9
		mLR : 0.999	8.95
	2/9/92	mLR : 0.999	9.25
		mLR : 0.999	9.65
Microcystis (Cropston)	20/8/92	no match	465
		Nod : 0.996	19.8
		mLR :0.999	31.1

IC_{50} (nM) : inhibition of phosphatase activity

REFERENCES

1   W.W. Carmichael,"Handbook of Natural Toxins", ed. A.T. Tu, Dekker, New York, 1988, Vol 3, p121.
2   I.R. Falconer, A.R. Jackson, J. Langley and M.T. Runnegar,"Liver pathology in mice in poisoning by the blue green alga *Microcystis aeruginosa*", J. Aust. Biol. Sci., 1981, 34, p179.
3   R.E. Honkanan, J. Zwiller, R.E. Moore, S.L. Daily, B.S. Khatra, M. Dukelow, and A.L. Boynton,"Characterization of microcystin-LR, a potent inhibitor of type 1 and 2A protein phosphatases", J. Biol. Chem., 1990, 265, p19401.
4   C. MacKintosh, K.A. Beattie, S. Klumpp, P. Cohen and G.A. Codd,"Cyanobacterial microcystin-LR is a potent and specific inhibitor of protein phosphatases 1 and 2A from both mammals and higher plants", FEBS Lett. 264, 1990, p187.

ACKNOWLEDGEMENT

This research was sponsored by Severn Trent Water, Birmingham, UK.
Their help and involvement, especially that of Les Markham and Helen Picket, is gratefully acknowledged.

Detection of Cyanobacterial (Blue-green Algal) Peptide Toxins by Protein Phosphatase Inhibition

Christine Edwards,[1] Linda A. Lawton,[1,2] and Geoffrey A. Codd[1]

[1]DEPARTMENT OF BIOLOGICAL SCIENCES, UNIVERSITY OF DUNDEE, DUNDEE DD1 4HN, UK

[2]CURRENT ADDRESS: SCHOOL OF APPLIED SCIENCES, ROBERT GORDON UNIVERSITY, ABERDEEN AB1 1HG, UK

1. INTRODUCTION

To date, microcystins are the most commonly detected cyanobacterial toxins on a world-wide basis and have been associated with animal poisonings and human health problems[1,2]. In surveys in the UK, hepatotoxicity accounted for >75% of samples found to be toxic (Codd and Beattie, unpublished results). These potent hepatotoxins are produced by the major bloom-forming genera including *Microcystis*, *Anabaena*, *Oscillatoria*, *Nostoc*, *Aphanizomenon* and *Gomphosphaeria*, which frequently occur in eutrophic and occasionally in oligotrophic freshwaters.

Microcystins are composed of three D-amino acids: alanine, erythro-β-methylaspartic acid and glutamic acid; two L-amino acids which are variable; plus two unusual amino acids. These are N-methyldehydroalanine and Adda (3-amino-9-methoxy-2,6,8-trimethyl-10-phenyl-4,6-dienoic acid) which has been shown to play an essential role in toxicity[3]. The cyanobacterium *Nodularia spumigena*, which is characteristic of brackish waters but can develop in freshwater, produces the cyclic pentapeptide nodularin, with properties similar to those of the microcystins.

In addition to causing acute hepatotoxicosis it has been demonstrated that these toxins are potent and specific inhibitors of eukaryotic protein phosphatases 1 and 2A (PP1 and PP2A)[4]. Recent work has indicated that these peptide toxins act as tumour promoters in a manner similar to the marine toxin okadaic acid, a known tumour promoter[5]. Protein phosphatases comprise several families of enzymes that catalyse the dephosphorylation of intracellular phosphoproteins thereby reversing the actions of protein kinases, which all comprise crucial mechanisms of control within the eukaryotic cell[6]. In order to assay the protein phosphatases inhibition, a substrate specific for PP1 and PP2A, phosphorylase *a*, was phosphorylated using ^{32}P labelled ATP. Dephosphorylation of the substrate by PP1 and PP2A results in release of ^{32}P as the substrate reverts to phosphorylase *b*.

The ability of microcystins and nodularin to inhibit protein phosphatases was exploited as a rapid method for detecting this class of toxins in cyanobacterial extracts and water samples from a reservoir known to support cyanobacterial blooms.

2 METHODS

Preparation of standards

Microcystin-LR and -LY were isolated from an acetic acid extract of *Microcystis aeruginosa*. Microcystin-RR was obtained from an extract of *Microcystis viridis* and nodularin was purified from a culture of *Nodularia spumigena*. All peptides were purified by a combination of solid phase extraction and HPLC. The peptides were hydrolysed and the constituent amino acids were quantified to provide accurate quantification.

Extraction of microcystins from cyanobacterial material

Freeze-dried cyanobacterial samples collected between 1989-1992 were extracted as described by Edwards *et al*[7].

Preparation of enzyme extract

Crude extracts of PP1 and PP2A were obtained from rat liver as described by Ingebritsen *et al*[8]. Rat liver was homogenised in 50 mM Tris-HCl containing 0.1 mM EGTA, 0.1% 2-mercaptoethanol and 250 mM mannitol (1(wt.):2(vol.)). The homogenate was filtered through two layers of muslin and centrifuged at 16,000 x g for 15 min. The supernatant was snap frozen in liquid nitrogen in aliquots of 200 μl and stored at -80°C.

Preparation of radiolabelled substrate

Radiolabelled phosphorylase was prepared using components provided in the "Protein Phosphatase Assay System" produced by Gibco BRL and [γ-^{32}P]ATP (3000/6000 Ci/mM). The kit provides sufficient substrate for 300 assays.

Protein phosphatase inhibition assay

The crude enzyme extract was equilibrated at 30°C and diluted to ensure that the protein phosphatase activity did not exceed 30% dephosphorylation in 10 min. Assays were performed as described in the Gibco BRL kit.

Quantification of microcystins by their inhibition of protein phosphatases

A range of concentrations of microcytsin-LR, -RR, -LY and nodularin were assayed for protein phosphatase inhibition to obtain a calibration range. Extracts of bloom material collected from Hanningfield and Abberton Reservoirs, Essex,

England in 1992, in which microcystins had previously been identified and quantified, were diluted by 10,000 or 100,000 to enable quantification.

Confirmation of identity of microcystins using liquid chromatography combined with protein phosphatase inhibition assay

An extract from Milton Loch, Dumfries and Galloway, Scotland, known to contain microcystins, was separated by reverse phase HPLC and fractions (1 ml) collected. Aliquots were removed and tested directly for protein phosphatase inhibition.

Detection of microcystins in water samples

Raw water samples from Hanningfield Reservoir were analysed for microcystins by HPLC on a weekly basis for four months. Samples were filtered, concentrated by C_{18} solid phase extraction and analysed by analytical HPLC with diode array detection. A 500 ml sample was concentrated to give a final volume of 200 μl and 10 μl of this was used to determine protein phosphatase inhibition.

3 RESULTS AND DISCUSSION

Detection of microcystins in cyanobacterial material

Microcystin-LR, -RR, -LY and nodularin inhibited protein phosphatases in the rat liver extract with IC_{50} values ranging from 10-90 pg.

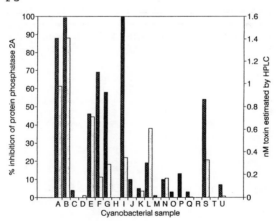

Figure 1. Analysis of microcystins in extracts of laboratory cultures and bloom material (A to U) by inhibition of purified PP2A (dark hatch) and HPLC (light hatch)

Twenty one samples of extracted cyanobacterial material were diluted 1/300,000 and tested for inhibition of purified protein phosphatase 2A (supplied by C. MacKintosh). All of these samples had previously been examined qualitatively and quantitatively by HPLC with diode array detection. All samples containing microcystins exhibited inhibition of phosphatase.

Quantification of microcystins in extracts of cyanobacterial blooms

Good correlation of the concentration of total microcystins between HPLC and protein phosphatase inhibition assay was obtained in all but one sample (Table 1). In most cases the concentration of microcystins was slighly higher when estimated by protein phosphatase inhibition; this was expected since the assay is several fold more sensitive than HPLC.

Table 1
Quantification of microcystins in extracts of cyanobacterial blooms by their inhibition of protein phosphatases

Source	Date Sampled	Concentration of microcystins	
		HPLC	ppase
		μg/mg dry wt.	
Abberton	26/6/92	1.7	1.8
Abberton	16/7/92	1.2	1.4
Abberton	9/7/92	0.4	0.4
Hanningfield	10/8/92	0.7	1.6
Abberton	30/7/92	0.2	0.3

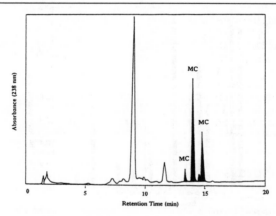

Figure 2. Elution profile of microcystins (MC) in a bloom extract from Milton Loch separated by reverse phase HPLC showing regions associated with protein phosphatase inhibition (shaded regions)

Confirmation of identity of microcystins using HPLC combined with protein phosphatase inhibition assay.

Protein phosphatase inhibition was only associated with fractions containing microcystins as shown in Figure 2.

Table 2
Analysis of microcystins in GF/C-filtered raw water from Hanningfield Reservoir

Date sampled	HPLC[a]	Protein phosphatase inhibition[b]
20/7/92	negative	positive (0.08)[c]
27/7/92	negative	positive (2.2)
17/8/92	negative	positive (2.7)
24/8/92	negative	positive (8.0)
7/9/92	negative	positive (3.0)
14/9/92	negative	positive (2.3)

a, detection of microcystins based on analytical HPLC with diode array detection

b, detection of microcystins based on inhibition of protein phosphatases

c, samples were diluted for quantification, amount shown ng/L

Analysis of microcystins in water samples

No microcystins were detected in water samples from Hanningfield Reservoir when analysed by HPLC (Table 2). However, all of the samples tested inhibited protein phosphatases. Several of these samples were diluted in order to quantify the presumptive microcystins, and concentrations determined were well below the limit of detection by HPLC (limits of detection are 100 ng/L and 0.4 ng/L for HPLC and protein phosphatase inhibition respectively).

4. CONCLUSIONS

Protein phosphatase inhibition clearly provides the basis of a rapid and sensitive assay for peptide hepatotoxins. It has two applications, firstly as a quick screen, replacing the often controversial mouse bioassay, and secondly as a method of confirmation, particularly in combination with HPLC, which has been successful for detecting okadaic acid and related diarrhetic shellfish toxins in shellfish meat[9]. However, it's usefulness for the detection of microcystins in water samples must be examined more thoroughly. From the data presented and other work, many samples inhibited the protein phosphatases when no microcystins were detected by HPLC. Although in some cases this may be due to low levels of microcystins, some of the positive results were obtained from water samples where cyanobacteria had not been present for six months. As recognition of the number of compounds that inhibit protein phosphatases is increasing, there is more risk of obtaining a false positive. Although some confirmation may be obtained on the identity of the inhibitory component(s) by examining their activity towards purified PP1 and PP2A, this is not practical for most laboratories. For water samples, it is most appropriate to use the combined HPLC-phosphatase assay approach.

ACKNOWLEDGEMENTS

We thank Professor P. Cohen and Dr C. MacKintosh for helpful interest and assistance in the early stages of this work, which has also been supported by Essex Water Company.

REFERENCES

1. Lawton, L.A. and Codd, G.A. *J. Inst. Wat. Environ. Management*, 1991, 5, 460.
2. Codd, G.A. and Beattie, K.A. *Pub. Health Lab. Ser. Micro. Digest*, 1991, 8, 82.
3. Rinehart, K.L., Harada, K., Namikoshi, M., Chen, C., Harvis, C.A., Munroe, M.H.G., Blunt, J.W., Mulligan, P.E., Beasley, V.R., Dahlem, A.M. and Carmichael, W.W. *J. Am. Chem. Soc.*, 1988, 110, 8557.
4. MacKintosh, C., Beattie, K.A., Klumpp, S., Cohen, P. and Codd, G.A. *FEBS Letts.*, 1990, 264, 187.
5. Nishiwaki-Matsushima, R., Ohta, T., Nishiwaki, S., Suganuma, M., Kohyama, K., Ishikawa, T., Carmichael, W.W. and Fujiki, H. *J. Cancer Res. Clin. Oncol.*, 1992, 118, 420.
6. Cohen, P. *Ann. Rev. Biochem.*, 1989, 58, 453.
7. Edwards, C., Lawton, L.A., Beattie, K.A., Codd, G.A., Pleasance, S. and Dear, G.J., *Rap. Comm. Mass Spec.*, 1993, 7, 714.
8. Ingebritsen, T.S., Stewart, A.A. and Cohen, P. *Eur. J. Biochem.*, 1983, 132, 297.
9. Holmes, C.F.B. *Toxicon* , 1991, 29, 469.

Investigation of the Solution Conformation of Microcystins-LR and -RR by High Field Nuclear Magnetic Resonance

Barbara Mulloy[1] and Robin Wait[2]

[1]LABORATORY OF MOLECULAR STRUCTURE, NATIONAL INSTITUTE FOR BIOLOGICAL STANDARDS AND CONTROL, BLANCHE LANE, SOUTH MIMMS, POTTERS BAR, HERTFORDSHIRE, UK

[2]DIVISION OF PATHOLOGY, CENTRE FOR APPLIED MICROBIOLOGY AND RESEARCH, PORTON DOWN, SALISBURY, WILTSHIRE, UK

1 INTRODUCTION.

The biological activities of the cyclic peptide toxins of cyanobacteria are highly sensitive to subtle chemical modifications [1,2], suggesting perhaps that a precise conformation is required for the maintenance of toxicity. Nuclear magnetic resonance (NMR) is the most powerful available technique for the high resolution study of three dimensional structures in solution, because it enables the determination of both through-bond and through-space connectivities between individual nuclei. We have therefore measured conformationally sensitive NMR parameters, principally nuclear Overhauser enhancements (nOes), and vicinal proton proton coupling constants ($^3J_{HH}$), which we have used to construct molecular models of the solution conformations of microcystins-LR and -RR.

2 MATERIALS AND METHODS.

Microcystins LR and RR were purchased from Calbiochem Ltd (Nottingham, UK), and (after verification of their structures by fast atom bombardment mass spectrometry), were used without further purification. Proton NMR spectra were recorded at 30 °C from 1 mM solutions of microcystin in 80% H_2O/20% D_2O, using a Varian Unity 500 MHz spectrometer. Solvent suppression was achieved by presaturation of the H_2O signal. One dimensional spectra were assigned by means of 2D COSY and TOCSY experiments. $^3J_{H,H}$ values were directly determined from the 1D spectra. Rotating frame nOes were identified from ROESY experiments, no attempt being made to quantify them. All two dimensional spectra were obtained using the manufacturer's standard pulse sequences. The mixing time for the TOCSY spectra was 80 ms, and the spin lock field about 10 kHz; for the ROESY spectra the mixing time was 120 ms, and the spin lock field was approximately 2 kHz.

The initial molecular model of microcystin-LR was assembled using the program Insight II (Biosym Technologies Inc), running on a Silicon Graphics IRIS 4D 310 GTX computer. Distance geometry calculations were performed using DGEOM (QCPE 590, Quantum Chemistry Program Exchange, Indiana University, Bloomington, IL, USA), and the resulting output structures were further refined with Insight II.

3 RESULTS AND DISCUSSION

The ^{1}H spectra of microcystins -LR and -RR were recorded in H_2O solution, enabling us to extend the assignments to amide protons; in other respects our assignments are in broad agreement with published data [2,3].

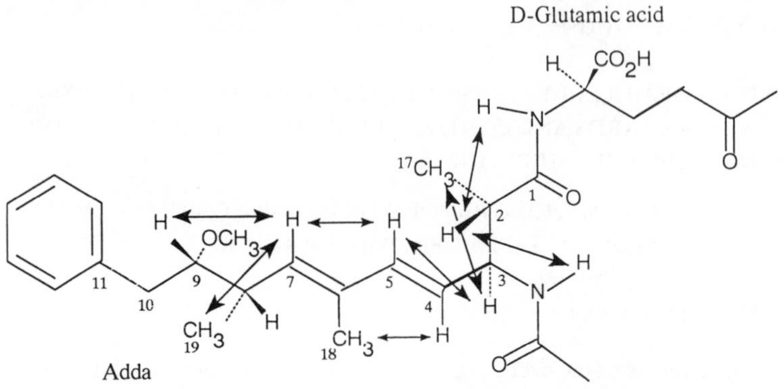

Figure 1 NOes used to constrain the Adda residue in the distance geometry calculations.

The NMR measurements were used to provide a set of dihedral angle and distance constraints for the program DGEOM. Short range distance constraints were obtained from observed nOes. These were only detected between protons in the same and immediately adjacent residues, suggesting that the molecules do not adopt conformations in which non contiguous regions are forced into close proximity. Torsional constraints were derived from $^3J_{HH}$ values, which are known to be a function of the dihedral angle between the coupled protons[4], splitting being maximal when the dihedral angle is 180 °. Both $^3J_{NH,C\alpha}$ and $^3J_{HC,CH}$ were used, the latter proving particularly useful in defining the amino acid sidechains, and the iso linked regions of the peptide backbone. The constraints applied to the Adda residue are summarized in Figure 1 and Table 1.

In the preliminary distance geometry study 16 separate dihedral angle constraints and 22 distance constraints were supplied to the program DGEOM. Corresponding calculations have not yet been performed for microcystin-RR, but since the NMR data were very similar in both compounds, it is likely that the conformations will prove to be broadly comparable. Figure 2 shows an overlay of microcystin-LR conformations which fulfilled DGEOM's acceptance criteria. Although there are some differences in the details of the ring structure, the overall appearance of most of the output structures is similar. Particularly striking is the orientations of the Adda and Arg sidechains, which extend in opposite directions above and below the plane of the ring. The conformation of the Adda residue is particularly well defined, and appears to be fairly rigid as far as C8. The smaller value of

Table 1 Dihedral constraints of the Adda residue used in the distance geometry calculations

Coupling	Value Hz	Dihedral Angle Degrees
$^3J_{NH,CH\alpha}$	9.4	~ 180
$^3J_{H2,H3}$	10.4	~ 180
$^3J_{H3,H4}$	9.0	~ 180
$^3J_{H7,H8}$	10	~ 180
$^3J_{H8,H9}$	6.8	undefined

$^3J_{H8,H9}$ however implies a greater degree of rotation about the C8-C9 bond. The suggestion[5], that the Adda sidechain adopts a U shaped conformation is difficult to reconcile with our experimental data.

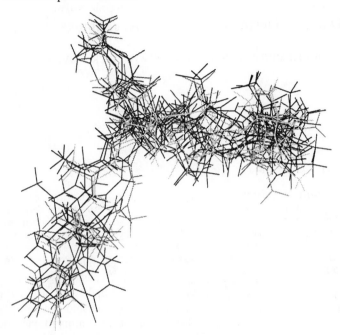

Figure 2 Overlay of acceptable conformations of microcystin-LR

It is likely that the relatively rigid and extended conformation adopted by the Adda sidechain is necessary for biological activity, since geometrical isomers at C7 are reported to be non toxic [2], and hydrogenation or ozonolysis[1] of the conjugated diene system likewise results in inactive products. By contrast, modifications which do not significantly alter the overall conformation of Adda, such as replacement of the C9 methoxyl substituent with a hydroxyl[6] or acetoxyl[7] group, cause no loss of toxicity.

REFERENCES

1. K. Sivonen, M. Namikoshi, W.R. Evans, M. Fardig, W.W. Carmichael, and K.L. Rinehart, Chem. Res. Toxicol. 1992, 5, 464.
2. K-I. Harada, K. Ogawa, K. Matsuura, H. Murata, M. Suzuki, M.F. Watanabe, Y. Itezono, and N. Nakayama, Chem. Res. Toxicol. 1990, 3, 473.
3. D.P. Botes, A.A. Tuinman, P.L. Wessels, C.C. Viljoen, H. Kruger, D.H. Williams, S. Santikarn, R.J. Smith, and S.J. Hammond, J. Chem. Soc. Perkin Trans. 1 1984, 2311.
4. A. Pardi, M. Billeter and K. Wüthricht, J. Mol. Biol, 1984, 180 741.
5. T. Lanaras, C.M. Cook, J.E. Eriksson, J.A. Meriluoto, and M. Hotokka, Toxicon. 1991, 29, 901.
6. M. Namikoshi, K.L. Rinehart, R. Sakai, R.R. Stotts, A.M. Dahlem, V.R. Beasley, W.W. Carmichael, and W.R. Evans, J. Org. Chem. 1992, 57, 866.
7. M. Namikoshi, K.L. Rinehart, R. Sakai, K. Sivonen, and W.W. Carmichael, J. Org. Chem. 1990, 55, 6135.

The Tandem Mass Spectrometry of Nodularin, Microcystins, and Other Cyclic Peptides

V. C. M. Dale, D. Despeyroux, and K. R. Jennings

DEPARTMENT OF CHEMISTRY, UNIVERSITY OF WARWICK, COVENTRY CV4 7AL, UK

1   INTRODUCTION

With the advent of more stringent water testing, the occurrence of algal species in our water systems has become a major concern to the water authorities and the consumer. Microcystin toxins are not destroyed by the chlorination, flocculation and filtration procedures of water treatment so the sporadic blooms of algae are a hazard to man.   The toxins produced are normally contained within the algal cells and are released only when the cell is damaged, either by physical rupture or more commonly on the death of the cell.   Thus, if such a bloom is present in a water reservoir and subsequently taken into the distribution system the probability of rupture and contamination would be high. Tandem mass spectrometry, together with an amino acid sequencer, has been used to characterise these toxins[1,2].

The Kratos "Concept" IIHH is designed to carry out the tandem mass spectrometry of large polyatomic ions, such as cyclic peptides.   The Collision Induced Dissociation(CID) of these peptides often yields little structural information. A cyclic peptide is more difficult to characterise than the analogous linear peptide, partly because of the problems in differentiating between the actual sequence and the retro sequence.   In addition, the fragment ions observed may arise from the ring opening of the cyclic $[M+H]^+$ at various sites, followed by losses of amino acid residues either from the N- or C-terminus of the resultant linear peptide.   For certain classes of peptides, useful additional information may be obtained from the CID spectrum of the cationated peptide.

In order to determine whether a more detailed structural characterisation of the blue-green algae toxins could be obtained, a study of the addition of each of the alkali metal cations, Li^+, Na^+, K^+, Rb^+ and Cs^+ to the peptide has been initiated.   The results obtained from the cationisation of nodularin are presented and discussed. Similar studies are being undertaken with both microcystin-LR and -RR.

2 RESULTS AND DISCUSSION

Ions observed in the CID spectra of [M+H]⁺, [M+2Li-H]⁺ and [M+3Li-2H]⁺, obtained using argon as the collision gas, are similar to those reported in the literature[3] for related compounds.

The CID spectra of the cyanobacterial standards show a number of side-chain fragmentations. The sequential loss of the Adda side-chain is observed, from the loss of the aromatic ring (A) through to the loss of the entire side-chain (F). An abundant ion at m/z 135 (G) results from loss of the tail portion of the Adda, PhCH₂CH(OMe); such a cleavage is seen for both the protonated and lithiated species. Accurate mass studies [3] of ions given by microcystin-LR and -RR have shown that ions at m/z 239 and 213 have the composition [CO+Glu+Mdha-H]⁺ and [Glu+Mdha+H]⁺ respectively. The corresponding ions at m/z 253 (H) and 227 (I) in the CID spectrum of [M+H]⁺ of nodularin are observed, whereas in the case of the lithium adduct ions only the ion at m/z 266 (J), [CO+Mdhb+Glu+2Li-H]⁺, is observed in the CID spectrum of [M+3Li-2H]⁺ . This, together with the lack of m/z 227 in the spectra of the lithiated species, suggests that at least one lithium co-ordination site is found in this region.

One might expect that a lithium ion would co-ordinate to the strongly basic side-chain of arginine. The presence of an ion at m/z 86 (K) in the spectra of the lithiated species indicates that this is not the case. This ion arises from the arginine tail. If a lithium cation binds in this region, the fragment would be shifted by six or seven mass units depending on whether a hydrogen is transferred to the neutral.

The spectrum of [M+3Li-2H]⁺ also shows a series of ions at m/z 665, 650, 636 and 622 (L-O) that arise from loss of m/z 135 together with the sequential loss of the arginine tail. The entire fragmentation sequence is not observed for [M+H]⁺ or [M+2Li-H]⁺. Each loses NH₂-C=NH together with the tail of the Adda chain of mass 135 hence yielding the ions of m/z 611 and 643 (M) respectively. This evidence suggests that the lithium ions co-ordinate to the ring of the peptide, probably to the Mdhb and Glu residues rather than to the side-chains; consequently these ions (L-O) arise from charge-remote fragmentations.

This study is being extended to other salts, such as the iodides, the hope being to obtain additional structural information. Different counter ions can alter the surface activity of the matrix/sample mixture and may also have some influence on the binding of the cation to the peptide. With the lithium iodide salt the most abundant adduct ion formed is the [M+Li]⁺ ion; CID of this ion will give an indication of where a single lithium ion binds to the peptide. A study of all the alkali metal iodide salts is being carried out.

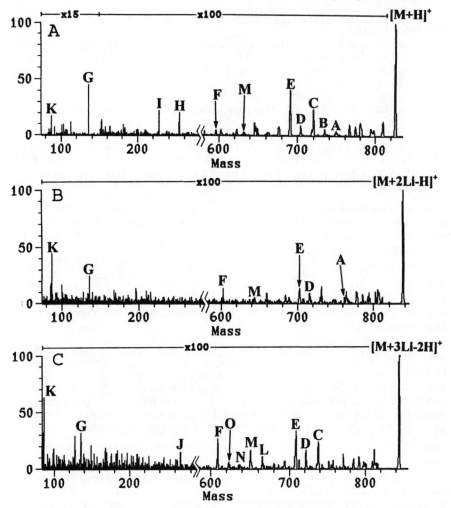

The CID spectra of A: [M+H]⁺, B: [M+2Li-H]⁺ and C: [M+3Li-2H]⁺
of nodularin. See text for explanation of nomenclature used.

3   REFERENCES

1. K.L.Rinehart, K.Harada, M.Namikoshi, C.Chen, C.A.Harvis, M.H.G.Munro,
   J.W.Blunt, P.E.Mulligan, V.R.Beasley, A.M.Dahlem and W.W.Carmichael,
   J.Am.Chem.Soc., 1988, 110, 8557.
2. K.Harada, K.Matsuura, M.Suzuki, M.F.Watanabe, S.Oishi, A.M.Dahlem,
   V.R.Beasley, and W.W.Carmichael, Toxicon, 1990, 28, 55.
3. M.Namikoshi, K.L.Rinehart, R.Sakai, R.R.Stotts, A.M.Dahlem,
   V.R.Beasley and W.W.Carmichael, J.Org.Chem., 1992, 57, 866.

4   ACKNOWLEDGMENTS

We are grateful to the University of Warwick and the Severn
Trent Water Authority for financial support during the
course of this work.

Subject Index